AF731047

DIGITAL TWINS

THE NEXT HUMAN (R)EVOLUTION THAT WILL DISRUPT THE FINANCIAL SERVICES INDUSTRY

JOHN BLICQ

DISCLAIMER

Digital Twins – The next Human (R)evolution
That Will Disrupt the Financial Services Industry

Copyright © 2021 by John Blicq

All rights reserved

First Edition

This book is the result of the author's 15 years professional experience in business management and corporate innovation, exploring the concept of Digital Twins and the implications for value chains in different industries. There is a strong emphasis on the financial services value chain.

The content of this book is mostly a theoretical exercise and by no means is it an attempt to predict the future. Instead, it leverages emerging trends and strong market signals which may materialise.

Care has been taken to confirm the accuracy of the information presented and describe generally sound and accepted concepts and techniques. However, the author shall not be held liable or responsible to any person or any entity with respect to any loss or incidental or consequential damages caused, or alleged to be caused, directly or indirectly by the information contained therein.

Published by:

Innovations
Accelerated

ACKNOWLEDGEMENTS

This book was written in Sydney, NSW, Australia.

I acknowledge the Traditional Owners of the Country in which this book was written, the Gadigal people of the Eora nation, and recognise their continuing connection to the land, waters and culture. I pay my respects to their Elders past, present and emerging.

I like to believe that the immense wisdom and connection to nature of the Australian Aboriginal inspired my writing.

For readers not familiar with the Australian Aboriginal history and culture, it is the longest and uninterrupted human culture that spans between 40,000 to 70,000 years.

For more information I encourage the reader to visit
www.reconciliation.org.au

GRATITUDE FIRST

I would like to first express my gratitude to a number of very special individuals who, without knowing it, have changed my life.

To my beautiful and immensely kind wife Karine who has shared my life for almost two decades and has played a fundamental role in shaping the human I am today. Thank you for your unconditional love and support over all those years and thank you for having given us our perfect daughter Emma. I love you both infinitely and I always will.

Thank you to the contributors of this book:

Brett Peppler, thank you for those years of mentoring. Thank you for those hundreds of hours of design sessions during which I witnessed your mastery of intelligence gathering and sense making. Above all thank you for your friendship.

Ben McMullen, thank you for your friendship, your dry sense of humour and your endless knowledge on digital transformation.

Dr Patrycja Slawuta, thank you for your friendship, thought leadership and your expertise in psychology which has helped me explore my own psyche. You are part of my family.

Edward Loh, thank you for your friendship, your vision, your wisdom and thank you for keeping me honest. Despite the distance you are close to my heart and I consider you a brother. Thank you for also being a beacon in the dark.

Thank you, Kirsten Dunlop. You believed in me and gave me a chance to explore the fascinating world of innovation in financial services. I wish the world had more leaders like you to drive positive change through innovation. We have a moral duty to prepare the minds for what is coming.

Thank you to all the members of the now disbanded Suncorp Group Strategic Innovation Team. You all changed my life. You opened my eyes on a vast range of possibilities. You gave me permission to explore and, more importantly, to be myself.

A special thank you to Ivan Vidovich, who has been one of the best leaders I ever worked for. Ivan you are a visionary and a wonderful human being who deeply cares about people.

Hannah Driscoll, your kindness and infinite wisdom have been invaluable support during our common tenure as executives.

Other incredible humans that I want to acknowledge are Will Soutar, Peta Karunaratne, Elana Robertson, Duncan Bone, Belinda Wheelan, Richard Riboni and Faith Byers. Thank you for your mentoring, your sense of humour and kindness. Thank you for these hundreds of hours of design workshops during which we created a possible different future. You are all very special people in my life.

Thank you for reading this book. I hope you find some elements that resonate with you and spark some insights.

PREFACE

Most people wonder "what's next"? We all intuitively know that human evolution has not ended, that many of us are already paying a high price for living and doing things in "faster, better, stronger" fashion as the ever-racing business world is asking us to. It's quite obvious that we, as humans, at this point cannot compete with machines when it comes to raw computational power, ability to search dataset with billions of data points or detect patterns in millions of pictures. This is where machine intelligence leverages human ingenuity by creating an exciting symbiosis that opens a new space of possibilities. Possibilities that have never existed before in the history of humankind.

What that exactly may look like - nobody really knows. Many speculate. Some worry. The brightest and smartest minds in this world are trying to imagine such futures. The futurists disagree - will it be utopia or dystopia? Or maybe just "business as usual" for us, average mortals stuck in the work, home, work endless loop...

If one believes the statement that "the future is already here, just unevenly distributed" then it's definitely worthwhile to see what are the elements of the future that already present today and deeply rooted that many people know what they are just by hearing the abbreviation: AI, ML, NLP, VR or AR. For some of us, these abbreviations sound as cryptic as names for products and services in another industry that touches the lives of most of us but also confuses most of us - financial services.

It is here, at this interesting intersection of two highly complex industries, advanced technology and financial services, where John Blicq and his book enters. With the French charm and in a relaxed Aussie manner John skillfully explains the so many not-so-easy-to-understand concepts: digital twins, blockchain, digital assets. Each chapter logically shows that our life already is augmented, distributed and highly digital and most of us already have some sort of polymorph virtual identity (or more) on platforms like LinkedIn, Facebook or by the fact we use multiple forms of avatars. The two, human life and the need to store and protect value, go hand in hand. Barter, shells, cows, gold, fiat currency, cryptocurrency - these are all ways humanity tries to store value of the things we care about to take care of the people we care about. John and his book brilliantly shows how emerging technologies are a natural evolution of this deeply ingrained human need.

Now, one question remains - is it an organic evolution or will it be a revolution in the way we live, love and behave. In order to answer that question you must read the book and actually decide for yourself. And then let me know...

Patrycja Slawuta, MA, PhD(c) Psych
patrycja@selfhackaton.com

TABLE OF CONTENTS

INTRODUCTION

The focus of this book is on the emerging value of digital assets, specifically on the concept of digital twins and how it will translate in the value chain of commoditised industries, especially financial services.

Although digital twins have been around for about two decades, quite simply, a digital twin is a virtual model of a process, product or service. This pairing of the virtual and physical worlds allows analysis of data and monitoring of systems to head off problems before they even occur. It prevents downtime, develops new opportunities, and even helps to plan for the future using simulations.

This book provides a glimpse into the impact of digital twins on financial services. The opportunity, impact and disruption to existing value chains within established and new financial services ventures. It provides strategies on how to leverage the power of digital twins to protect and grow revenue.

This emerging capability will change the way we approach a multitude of foundational capabilities. From risk management, products and services development, technology stack build and consumption, data, customer relationship, workforce management, physical footprint, no part of the value chain will be left untouched.

In these pages we explore the contemporary utopia and its impact on the financial industry, and why long-term thinking and innovation matters. In the new world, different forms of assets emerge and shifts what our generation will experience. Data will fuel our digital twins using technologies such as artificial intelligence, machine and deep learning, augmented reality and, most exciting of all, quantum computing.

As new innovations become reality, it will shift the traditional notion of assets within the financial services industry and your intellectual property becomes a valuable digital asset.

This book is for company directors, senior executives, enterprise architects, technologists, innovators, venture capitalists, consulting and advisory, economists, forecasters, human resources, marketers, lawyers, government, policy makers, and their reason to exist; customers (humans and digital humans).

If you work in financial services or are a client of it, chances are that now is the time to prepare yourself for the next evolution of this industry.

WHY LONG-TERM THINKING AND INNOVATION MATTERS

Having worked in innovation in financial services, often senior executives use the term "utopic" to describe the work we do and the value propositions we develop.

They are right because in a corporate world the reality is bounded by profit and losses, assets and liabilities, and time is counted on a 12 month horizon divided in weeks, months, quarters, semesters and financial year.

Increased visibility and scrutiny from Boards, with increased personal accountability to protect the company and its stakeholders (investors, people, regulators, community), mean management have an increased focus on value generation.

Senior executives' reality is grounded and timebound to deliver short-term outcomes. It is the Board responsibility to influence management to prepare their organisation for beyond the near term to build a sustainable future.

Innovators and strategists have a duty to work on a different time horizon. They must see the world with different lenses, they must pick up signals of emerging trends which will impact the value chain of the company they serve.

For an innovator or a strategist, thinking 12 months (short term) is simply not possible. It would anchor the thinking in the finite game where competitors remain the same. Where customers, society and collective consciousness are static. A 12 month world doesn't change and there is little to learn about, nothing to adapt to.

Instead, innovators and strategists project their thinking in long-term horizons or the infinite game. Once you break free from the constraints of a static environment endless possibilities emerge. The work of the innovators is to identify and make sense of the trends which will impact the future of their organisation.

The theory highlighted in this book would not be possible in 12 month anchored thinking. We will see that digital twins will likely become multi-generational digital assets.

As innovators, our work can be described as utopic but here are some examples of utopia which have changed the world we live in and left well-established incumbents into business school cases of failure to adapt:

- The utopia of a spheric Earth (Copernicus 1543) challenging the all mighty Catholic Church dogma and opening the minds for scientific exploration. (We can confidently say that Copernicus was right too early. It's saddening to witness that, despite undeniable scientific evidence; 500 years later some people are still convinced that Earth is flat.)

- The utopia of spreading knowledge across the world. Leading to the invention of the printing machine (Gutenberg 1439).
- The utopia of finding a shorter and quicker route to the Indies (Christopher Columbus 1492) leading to the discovery of a vast new continent and unknown advanced civilisations.
- The utopia of the enlightenment offering mind space for a world in where people would be free to choose their beliefs and experience different forms of governance than total monarchy. This led the people of France to their Revolution, opening a pathway to modern democracies.
- Closer to our time, the utopia of sending people into space and to the Moon; propelling research and science leading to life changing discoveries and pushing the boundaries of our earthly reality. At the time of writing this book, NASA had landed its rover "Perseverance" on Mars with the mission to find traces of former life on the red planet.

These few examples demonstrate why utopia matters and how it can change the world. Let us have a quick look at what was utopia only a decade ago is now changing the reality of the financial services value chain.

A CONTEMPORARY UTOPIA IMPACTS FINANCIAL SERVICES INDUSTRY

In this chapter we will focus on contemporary technological and societal trends impacting the financial services industry. These trends are now well established and likely here to stay. The common denominator is that they all started with a clear utopia in mind.

- Blockchain
- Autonomous vehicles
- Connected environments
- Peer-to-peer exchange
- The gig economy

BLOCKCHAIN: A WORLD WITHOUT BANKS

The utopia of a world without banks led to the development of cryptocurrencies based on blockchain technology. What started after the Great Financial Crisis of 2008 as an underground utopic developers' dream is now traded at a unit price of $US51,500 on the day of writing (early January 2021) and could be a new form of gold. (If you are interested in understanding what blockchain and cryptocurrencies are and how it works, I recommend *Blockchain: Understand Blockchain in a Day* by Steve Gold). Blockchain is a decentralised protocol that enables an immutable record (a single source of truth). Technically it is un-hackable. The concept emerged after the GFC when there was little confidence in the financial system. At the time it was designed as an alternative system to the opacity of financial transactions. Bitcoin (one of many) was designed to replace the transaction layer of the banking system. (For a deep dive into payments and banking I recommend *Payments and Banking in Australia: from coin to Cryptocurrency* by Nikesh Lalchandani).

The information is cut in pieces (hash) and stored across a distributed network of participants. It is encrypted in a way that all participants must be hacked at the same time to reconstitute the information. There are multiple forms of protocols; both public and private. In a public blockchain the information is available publicly, in a private one, only to the participants. The insurance value chain can be massively disrupted and or enhanced using blockchain technology.

Insurance companies apply risk-based pricing on their static knowledge of certain asset classes. To do so they rely on external data (weather flooding area/earthquake area/bushfire area/accident frequency, etc) but also data correlated to the build of an asset (material/maintenance/etc). Having a single view of an asset is a key factor to price risk accurately; however, a true single view is difficult to achieve in practice. For example, those records could be transferable from one owner to the next and would help insurers identifying the risk profile of an asset.

Having access to this sort of technology helping to prove provenance and ownership over time could have a massive impact on fraud mitigation.

From a commercial perspective, blockchain can enable "smart contracts" where transactions (exchange of ownership/liability/payments/claims) can be automatically triggered by measurable events and predefined thresholds.

This could also enable the development of "invisible insurance" products (e.g. flight delays: your flight is delayed by more than 30 minutes, you are automatically compensated before you even land and you don't have to go through a claim process). Identity insurance could also be developed to protect from identity and reputation theft.

Why this discussion of blockchain? Blockchain is a core technology that supports the concept of digital twins – the subject of this book.

Blockchain is powerful in proving provenance of an asset (physical or digital) it could prove components of a digital twin such as digital identity and digital reputation. It could also prove components of the identity of a physical asset, for example, virtual records of maintenance (digital logbook) of a car, property, boat, warehouse, plant.

AUTONOMOUS VEHICLES: A WORLD WITHOUT DRIVERS

The utopia of flying cars now translates into autonomous vehicles (AV) that will dramatically reduce the number of fatalities on the road by removing the main risk factor: the driver. This will significantly affect the insurance value chain. On a 5+ years horizon, AVs will scale and likely start as a mobility service provided by taxi companies, ride hailing/sharing companies such as Uber, Lyft, Didi or Grab. They will be on demand and allow the passenger to focus on anything but driving.

As a side note, flying cars or drones are not far down the track, companies like Airbus and Hyundai are working on flying vehicles destined for urban mobility.

Before it becomes the norm, human-driven vehicles and AVs will cohabit on roads and crash into one another (likely the human will be at fault). This will change the way risk estimation and underwriting are done. As expensive technologies make their way into the design of cars, they will become more expensive to repair. Insurance premiums will have to adjust to a point where AVs may simply not be insurable. It is expected the first generations of fully automatic vehicles will cost hundreds of thousands of dollars. Insuring an AV for the average person might simply not be affordable.

Under our current model a bank wants to sell you a loan so you can own a car, an insurer wants to sell you a policy in case damage occurs to your asset so you can fix or replace. In most cases cars are used for short trips and they are parked somewhere, unutilised for 90%+ of the time. During that time, you are reimbursing a loan and paying an insurance premium for a depreciating asset that is not exposed to major coverable risks. In short, a regular car owner over consumes for a limited service rendered by the ownership model of a car. Financial services have geared to support this ownership model. Insurers must be very efficient at fixing cars; some have an integrated model where they own repairer networks to leverage economies of scales and use them as an adjacent source of revenue.

Let us now shift the assumption that "people will want to own a means of transport" to "people will want to use mobility solutions". A whole new world emerges and threatens the traditional financial services value chain. Part of the world of mobility encompasses public transport (buses, trains, trams), ride hailing and ride sharing (taxi, Uber, Lyft, Grab), and micro-mobility options (free floats of electric bikes, pushbikes).

As urbanisation seems to be an unstoppable trend, there will be a competition for living spaces. In most urbanised areas the space dedicated to cars, in the forms of roads, car parks public or private, constitute up to 30% of the available land. Mobility services and AVs are likely to reduce the number of cars on the street. Increasingly, major cities like London and Paris, not mentioning Singapore, have open public policies going against the car ownership model. As the number of cars reduce on the road, the revenue pool available to banks and insurers shrinks.

Saudi Arabia's Crown Prince announced in early 2021 the building of a totally new 170 km long city called "The Line". It will be 100% pedestrian with a network of hyper-speed underground transportation enabling an end-to-end journey of the city in about 20 minutes. In that new type of city, a bank or an insurer cannot derive revenue from a traditional financial product associated to the car ownership model.

Another adjacent effect of removing cars from the streets will be the creation of more living spaces to build homes, parks, leisure centres, cultural centres etc. We can expect a significant impact on the prices of real estate if suddenly 30% of an urban area becomes available for development. In that scenario, the scarcity economy the real estate industry leverages will be challenged. This leads us to another challenge for the financial services industry value chain. The challenge of connected environments.

CONNECTED ENVIRONMENTS: AN AUTOMATED WORLD

The utopia of automated and harmonious living spaces led to developing modern cities: immensely complex and vast stretches of land home to millions of inhabitants where food and utilities are abundant. Something as trivial as turning the water tap on in a top floor apartment and having drinkable water almost a kilometre above the ground was science fiction before the first high rise buildings emerged in the US in the early 1900s thanks to innovation in steel infrastructure and elevators. We commonly now talk to devices in our homes; to clean living spaces, to control the temperature, the light and various appliances. Light bulbs use light beams to enable Wi-Fi to our connected devices. As we move deeper in that hyper-connected world, the risk profile of what remains the most expensive asset an individual can own (a home) changes. Financial services need to understand that changing nature. Traditionally, financial services establish a property risk profile and value of a property using conventional data points such as: location, construction materials, size, age, external environmental factors having impacted the location over the years.

Location generates an idea of the inherent value based on market dynamics. Location also gives an indicator of external risk exposure such as flood zones, exposure to weather events such as cyclones or acts of God such as earthquakes.

Construction materials and the age of the property gives an indication of the trade-off between obsolescence and resilience to external events.

Based on this data the financial services industry can provide the aspiring owner with a mortgage and property insurance.

But this conventional approach to risk assessment disregards the data and behaviour generated from within the property. A water leak can damage the structure and remain undetected for decades. A humid environment coupled with a poorly-ventilated space can lead to the growth of mould/mildew that can be harmful and, in certain cases, deadly to the occupants. A connected surveillance system can give instant awareness of burglary attempts.

Growing connected in our home environments will enable the management and mitigation of the inherent risks occurring within the property. Players like Google through Google Nest, Amazon through Alexa and Ring, and Apple through Home Pods will be well positioned to capture these critical "unconventional" data points and change the way the financial services industry underwrite risks.

As the science of construction material engineering progresses, material property will change and provide more resilient properties. It is easy to imagine roof tiles that becomes waterproof when it rains or more resistant to shock when a hailstorm occurs. We can also easily imagine walls, doors and garage doors that become waterproof when a flood is predicted.

From an asset value perspective, the price is driven by stock availability, willingness and capacity to buy.

In major cities where potential buyers compete for home ownership, prices skyrocket, and it becomes more difficult for first-time buyers to enter the property market. This trend is accelerated by mainstream central bank policy of low interest rates (quantitative easing).

The development of 3D printing technology has made tremendous progress over the past decade and it is now possible to print a home in 48 hours for about $50,000. Entire cities are being printed in China to support massive urban expansion. Even shipping containers are being used to create modular homes at a fraction of the cost of a traditional house.

Connected, smart, resilient environments present a serious challenge for banks and insurance companies that derive a significant part of their revenue from selling mortgages and home, contents and mortgage insurance.

The reason why those highly profitable products exist is because owning a house is the most expensive and valuable asset a regular individual can aspire to. It is a nest for growing a family, it is a safe place, and it is a pathway to financial freedom once the property is owned outright.

Should solutions like 3D printing make home ownership more accessible, this will significantly threaten the financial services revenue stream. Indeed, rather than financing a $500,000 property over 30 years, people will be able to print a house for $50,000. The loss of revenue in the form of interest will be significant for banks to absorb. Finding alternative revenue streams will be paramount to survival of the retail banking divisions.

Coupled with a changing nature of risk understanding, now made available by connected environments, banks and insurance companies have a serious challenge ahead.

PEER-TO-PEER EXCHANGE: A DISINTERMEDIATED WORLD

Another utopia that challenges the financial services industry value chain, consists in enabling people to unlock inherent value by monetising their own assets in a peer-to-peer (P2P) way.

Platform business models have been key enabler of that vision. A platform is a digital double-sided marketplace where offer and demand meet through orchestration.

PLATFORM BUSINESS MODELS

The most successful of all platform business models is Amazon. This company aspires to become the provider of everything. Over time Amazon, by being first and foremost a customer obsessed (they coined the term) technology company has successfully disrupted many industries. Its ability to spot a weakness in a value chain and provide a better customer experience barely has any equivalent. By nature, a platform business model is a marketplace. There is nothing new about the notion of a marketplace. It probably is the oldest form of trading we know of. What is revolutionary is that it is a "digital" marketplace, capable of scaling almost instantly across the world to serve the needs of any of its 7 billion users.

Amazon started with selling online and delivering books because it is one of the most complex products to manage and ship. There is a plethora of books across uncountable topics, written in hundreds of languages. If Amazon could be successful with books, they would succeed at anything. It forced Amazon to be the world's best across many capabilities: technology, customer (providers and end-users) relationships, pricing, logistics, etc.

One of Amazon's core capabilities is cloud computing. Amazon Web Services (AWS) is now one of its most profitable sources of revenue. Cloud computing consists of enabling third parties to access key infrastructure and software. In Amazon, instances they opened their massive data centres and storage centres to offer computing power to the masses. Because they use their own technology to expand their business, AWS customers know they will benefit from the world's best technology available. If it is good enough for Amazon, it is probably good enough for most of the businesses across the world. The economics also works well. To use cloud computing services, it costs a fraction of the resources needed to maintain it onsite. A business must focus on its core and let technology be the problem of world class experts.

The monetisation of a platform businesses mainly comes from a "clip of the ticket" model, whereby a portion of the cost associated with a sale goes to the facilitator (Amazon in this instance). The more the offering widens, the more it attracts consumers. The compounding effect of that model has made Jeff Bezos, Amazon's founder, the richest man on the planet. To this day there are no limits to the expansion of this model and many companies are trying to replicate this model by creating ecosystems of services relevant to their targeted customer base.

THE GIG ECONOMY: A WORLD WITHOUT EMPLOYERS

Platforms such as Airbnb and Uber, only to name the most commonly known, enable anyone to "sweat" their personal assets. Anyone with a car can turn into an Uber driver and can either use this model to create an additional source of revenue to complement a full-time job. Or use this means of revenue generation as a main income stream. Airbnb enables any landlord with a spare room to monetise it by welcoming strangers into their house for a night. Both those companies started with the concept of using technology to enable people to unlock under-utilised assets. They found a weakness in the value chain of the industry they were aiming at disrupting e.g. the hotel and taxi industries. They are overcoming regulatory challenges by having a proposition so compelling that users (on both sides of the platform) quickly adopted it and created a habit.

The taxi industry thrived by acquiring expensive licences to operate in cities. The limited amount of available licences created duopolistic environment where only a few taxi companies could operate with regulated tariffs because of the high cost of entry. This created scarcity in the offering while urban populations were increasing rapidly. Ride hailing apps have flourished and broke the barrier to entry by enabling any car owner to fulfil an increasing demand. All this at the expanse of taxi companies who were now facing unfair competition. This model has been so successful that it has enabled long time unemployed to get back to work. This now creates new challenges by breaking a new barrier — the employment one. It is a fair to ask whether an Uber driver is an employee or a service provider. That raises some significant tax-related questions.

Airbnb uses a similar model but applied to the tourism industry. Again, an unfair competitive advantage has enabled Airbnb to become the world's largest provider of rooms without owning a single hotel.

Although Airbnb and its competitors are enabling property owners to generate extra income, in certain cities it has become a serious threat to public monopolies. Cities like Paris or London, which derive non negligible sources of funding from tourist tax on accommodation are now limiting or making it more difficult to participate (as a provider) in these services. The financial incentive for a property owner to use the Airbnb platforms was so great that it has fuelled asset value spikes in certain cities. Apartments were bought with the sole purpose of short-term rental.

In densely populated areas where the cost of accessing property (rental or ownership) is already high, services like Airbnb infuriate constituents and pose challenges for local authorities.

Whether we adopt those platforms or refuse to, they are here to stay. Regulators will have to find a way to cope with the anger it generates across multiple parties, but they are widely adopted, and they will continue grow. Those platforms also fuel adjacent industries. The service industry: ride hailing platforms have led to the creation of new ways of consuming mobility and catering for it. Companies have specialised in providing drivers with packaged services (cars + insurance + fuel) for a weekly fee, lowering even more the barrier to entry. Now anyone with time on their hands can generate income through those platforms. Short-term stay platforms have led to the creation of dedicated cleaning, concierge and catering services. These companies require labour to fulfil new needs.

These platforms also force financial services to rethink parts of their service offerings. They force insurers to rethink the notion of "risk" and how to price it, which is a fundamental capability of the insurance industry. For example, an Uber driver is spending more time on the road than an average driver. That person is not necessarily a road professional like a taxi driver and that person transports other non-related individuals. These factors significantly increase the nature of risk and exposure. On a similar logic, an Airbnb host welcome strangers into their property, which also changes the nature of risk and therefore premium.

THE GIG AND COLLABORATIVE ECONOMY

Platforms like Freelancer.com, Upwork.com and others enable freelance workers to break free from the traditional employer-employee paradigm in which employees monetise their time for the sole benefit of an employer. The collaborative and gig economy gives people the option to take control over their employment path and lifestyle. Some use the gig economy as their primary source of income where other monetise their skills to complement another recurring source of revenue. It also provides advantages for consumers of those services to access highly qualified professionals in a cost effective manner. Indeed, the gig economy shortcuts any form of intermediation such as creative agencies and recruitment agencies.

I have just described some examples of what utopia looked like in a different period of time across human history. Some utopias are very recent and have become reality in matter of a decade or two.

Let us now talk about the utopia of digital twins and do a thought experiment for a moment.

THOUGHT EXPERIMENT – TOMORROW'S WORLD

It is the morning of July 14, 2030, at 6.30 am your smart implant (they have replaced smartphones about 5 years ago) gently buzzes your cerebrum (the part of the brain that controls senses) to wake you from a restful night. An electric signal is induced and makes your eardrums hear a pleasant nature soundscape. After 5 minutes of this music, you slowly bring your conscious back to your body. You get out of bed and your preferred selected information is projected onto your optic nerves. You received a message from a relative living across the world, weather will be great, and you have an appointment with your digital doctor at 1 pm.

You also have 126 meetings today. Your attendance is required at three of them. Your intellectual property licence renewal with your main client at 10 am; your IP lawyer at 12 pm to seal the new contract and your banker at 3 pm. Be ready for this one as you are likely to get a new home loan by leveraging the financial performance of your enhanced digital twin. You have trained it for months with new IP downloads to generate 15.7% more income with your targeted audience. You will be handling your negotiation with your client's counterpart and their bot lawyer. It is an important conversation, but your results have been outstanding, your IP has delivered output that ranked in the top 5% of their digital employees. No need to worry. Your digital twin will be your wingman. It will handle the conversation with the bot lawyer and lead the client presentation on a different channel. In fact, the contractual key terms have already been pre-agreed two years ago and your compliance in meeting the service level agreement is recorded on your private blockchain.

You get in the shower for a fresh start. Meanwhile you ask your digital twin for an update of the worked achieved last night. There were 36 transactions for a total of $2,568 — 70% with other digital twins, 27% with digital humans and 3% with humans. Not bad for a restful night. Interestingly you note that your digital twin has increased transactions with Digital Humans located in France by 7.33% over the past three days. Perhaps you should reinforce your digital presence in French speaking countries. You ask your digital twin to ensure it has downloaded the latest update in his program in French. You also ask him to increase marketing expenses on your French professional holographic demo.

Now that you are refreshed, it is time for your morning walk on the beach with the family. Don't forget to put sunscreen on. UVs are high today and don't be late for your human meeting in three hours.

SCIENCE FICTION OR REALITY?

What I describe sounds like science fiction but the likelihood of this scenario arising may surprise you. William Gibson is often quoted in the innovation community, he said "The future is already here, it's simply not evenly distributed".

All the necessary external conditions (NECs) exist for this scenario to occur. To the best of my knowledge, they are just not yet connected together.

Will this raise many philosophical, ethical, medical, legal questions? Of course, it will, but it's not the first time that our species enhances itself. Wearing prescription glasses, hearing devices, replacing a broken tooth, having a pacemaker installed, using a prosthesis to replace a lost limb, undertaking surgery to have an artificial heart implanted are all forms of human enhancement.

In late 2020, the European Union (EU) approved the launch to market for artificial hearts from the French company Carmat[1]. All of the above previous examples are forms of bionic humans. Our generation does not even consider this enhancement bionic because it became so common. Imagine the reaction of our grandparents or great grandparents if we could have told them their wounds and

---

1 https://innovationorigins.com/green-light-for-european-artificial-heart/

permanent injuries from experiencing two world wars could be mended by this sort of technology. They would probably consider it impossible. Our grandchildren or great grandchildren are likely to have the same feeling about us when it comes to digital twins.

Some core components of the thought experiment we went through earlier in this book are already available, it's a matter of time for these technologies to scale and reach the masses. We are going to go through it in detail.

But first and foremost, here are some reasons why this digital twin scenario could well apply to a significant portion of the developed world.

The western world, at the turn of the early 90s made a strategic decision to transform their economies from being heavily driven by manufacturing to becoming more oriented toward services. The environment was very favourable for that shift, well-educated population, higher level of potential growth for the services industry, rising manufacturing capability and cheap labour from "Asian tigers".

We have reached a point where modern western economies rely on services for more than 70% of their GDP. According to the European Central Bank, the percentage of GDP driven by the services industry, in 2019, was 73.7% for the EU and 79.8% for the US. In service-oriented economies, workers no longer deliver physical and tangible outcomes in the form of products. Rather, workers monetise their knowledge, experience and self-identity to their employers. It is a form of exclusive intellectual property licencing. Each worker is unique, and their work is the result of their identity such as personal (values, beliefs, education) and professional qualifications and experience.

For a significant period between the 1990s and the mid-2000s, collaborative technologies were not sufficiently advanced to enable remote working at scale.

In the current context, accelerated by a global pandemic (COVID-19), the professional world has had no choice but to embrace remote working, collaborative tools and, more importantly, digital transformation.

Remote working is likely to become embedded into our habits and perhaps it will even become the norm. Should this materialise, it will reshape urban dynamics. But more importantly, it will force rethinking the notion of location and productivity. This represents a shift in management style requirements. If physical presence was associated with control and productivity, COVID-19 probably challenges that paradigm.

As we transitioned overnight from the routine of commuting and clocking in/out to full remote working and lockdowns, we discovered a new form of work/freedom addiction.

In terms of the benefits, workers have saved a significant amount of time in avoiding the commute to work and enjoyed more personal disposable times.

But on the downside, workers have experienced an invasion of their private space (home) by work. This physical amalgamation and urging need to be connected has driven people to work extended hours. This has likely led to a surge of productivity for their employers. According to a Boston Consulting Group[2] survey of 12,000 professionals in various countries, 75% of employees said that during the first few months of the pandemic they have been able to maintain or improve productivity on their individual tasks.

---

[2] https://www.bcg.com/en-au/publications/2020/valuable-productivity-gains-covid-19

It is difficult to now consider a return to the previous world. A hybrid model of mixed remote and onsite working is likely to prevail, but a scenario of full remote working isn't unrealistic in the near future.

This book talks largely about the next human evolution which is being powered by technology but first let's try to put this in the context of human evolution.

A QUICK HISTORY OF EVOLUTION

HOW DID A WEAK SPECIES RISE TO THE TOP OF THE FOOD CHAIN?

We have taken for granted our long-acquired status of super predator. In fact, most of us probably don't recognise themselves as such. It's been for generations that we are on top of the food chain, without any real threat to our existence except for the human created ones.

A rather fascinating question is: How did that happen?

How did a rather weak biped species like Homo sapiens rise to the top in a quite short period of time?

Our ancestors were equipped with a "hardware" comparable to the one we use today. Weak in many aspects such as limbs and the senses (smell, sight, earing). We were not even granted remarkable physical abilities. A chimpanzee, despite its smaller size is far stronger than a regular human. One can argue that they would outsmart most humans.

We are indeed so unremarkable against top predators that, according to Greek mythology, Prometheus (a Titan, punished by Zeus and forced to be his operative force to shape and gift all living species) took pity of humankind and stole fire from the Gods to help mankind survive.

Without fire we were at the mercy of cold weather and attacks from physically stronger animals.

Weak we are but yet we have survived, thrived and became a threat to our own existence. What makes us unique is two key attributes, great endurance, and a relatively large brain for an animal our size.

This brain and consciousness have evolved over time. From the very beginning we have compensated our limitation by an unmatched ability to innovate and shape our physical reality. Over a rather short period of time (10,000 years) we have been the only species to use widely accessible resources and transform them to our advantage.

This ability to think in context, hypothesise, test, learn and adapt has given us our status of total dominance.

That was before the widespread of abundant and cheap computing power. This leads us to a significant dilemma.

HOW DID OUR "WET WARE" OUTPACE OUR "HARDWARE"?

Over the past few thousand years, our "wet ware" (brain and consciousness) has evolved in a quite linear fashion while our "hardware" (body) remained unchanged. Surely the past hundred years have accelerated that trend of wet ware evolution. It is said that an average westerner is exposed every day to as much information as a person from the 15th Century was exposed to in their entire life. But it's far from being enough to help Homo sapiens comprehend another fast evolving being —technology.

We are still the by-product of the environment we evolve in. For westerners, like myself, our identity is forged in Judeo-Christian dogma, influenced by the philosophy of the enlightenment, the scientific approach of Descartes, the biases induced by our education (parental and scholar) and an enhanced appetite for consumerism. Past those details we are pretty much similar beings that our ancestors were 10,000 years ago.

On the other hand, technology has become, in half a century, abundant, cheap and incredibly powerful. While we evolve in a linear fashion, technology evolves exponentially. According to Moore's law[3] the number of transistors per silicon chip doubles every year. In other terms, if technology was competing with Homo sapiens for Earth supremacy, we would have lost the battle a long time ago.

Instead, every day, we walk around with super computers in our pockets. Each of our smartphones have more computing power than the Control Centre which enabled our species to land on the Moon in 1969. Instead, technology enables us to be smarter, more productive, more conscious but also more biased and more addicted. We are physiologically and psychologically not equipped to win that race against technology. We can only hope that our collective wisdom will lead us to use technology to enhance our abilities and not hinder our evolution.

With my own biases I did choose the optimistic camp, believing that we will use technology as an extraordinary medium to enhance our species, fight world challenges such as climate change, reduce inequalities among our own species but also to prevent damage to other species, create wealth and abundant, sustainable resources through relentless delivery of innovation.

---

3 https://www.britannica.com/technology/Moores-law

Regardless of your opinion on this utopic vision what is certain is that the exponential ability that technology has to create outcomes, leads to a paradigm shift.

Next, we explore the different asset classes at play and how digital is likely to overrun physical in terms of value perception.

INTRODUCTION OF ASSET CLASSES AT PLAY

(PHYSICAL VS DIGITAL VS PHYGITAL) – INDUSTRIAL ECONOMY VS DIGITAL ECONOMY

In this section we explore different forms of assets and what shift our generation will experience.

From time immemorial we have valued physical assets. It has been a very efficient way to trade, swap, borrow and loan. After all a physical asset can be used to fulfil a physical or aspirational need. It can be transferred and seized; hence, its intrinsic value can be used to liquidate a risk position.

Since the Industrial Revolution, we have made a societal decision to use physical metrics to count the output or wealth created by an entity (e.g. an institution, country).

Entire value chains have been organised to produce and fulfil needs articulated around physical assets.

How many tonnes of a raw product was generated by a factory? How many tonnes of that raw product were shipped? How many instances of transformation and value have been added? How many finished products were shipped to clients' warehouses? How many units have been sold to customers?

Companies and individuals count their wealth in terms of physical assets (stock, production lines, properties, motor vehicles, number of units of currency in a bank account, number of shares in a portfolio, etc). Most assets have physical and tangible underlying value.

But technology enables the (fast) creation of digital assets.

Digital assets are difficult to properly define. A simple way to approach this concept is by thinking of any kind of content, stored digitally, that can be monetised. A distinction can be made between established versus emerging digital assets.

A database, a data lake, a sensorial content (photo, video, music) are forms of established digital assets. They can be transferred, licenced and easily monetised.

In terms of emerging digital asset class, a digital wallet containing crypto currencies, a digital form of identity, a digital reputation, a social media account are forms of digital assets.

As most of us consume digital content, think surfing the internet, we all create a growing digital footprint. This footprint is used by technology giants to create personas. They use this emerging form of digital asset to provide consumers with tailored offerings. In that instance, the digital asset is monetised by the aggregators.

To understand the potential power of digital assets, let us run a thought experiment. Let us pretend you bought a $500,000 property in one of the world's capital cities in 2008 – 2010 (e.g., NYC, Tokyo, London, Paris). Should you be liquidating this physical asset today

(2021) you would probably have made a very decent capital gain (+25% minimum). We are all familiar with this sort of scenario.

Now let us pretend that in the same period you had invested $100 in a cryptocurrency such as Bitcoin. If you wanted to liquidate that digital asset today you would be a billionaire.

Now let's assume that you hold these two different asset classes (a property and a crypto wallet), which one do you want to absolutely protect? Of course, the answer is both. But with your digital asset, you would struggle to find a bank or an insurer equipped to provide you with a fit for purpose protection or financing mechanism. Should you need to only protect one asset class, any sane person would protect the digital asset.

This dilemma illustrates the concept of digital assets as an emerging asset class. The consequences for many industries (especially financial services) are enormous because the industry has optimised its value chain to provide services for individuals (people and organisations) to acquire physical assets.

If we shift the paradigm from "people are going to value physical assets" to "people are going to value digital assets", the traditional financial services value chain collapses and non-traditional competitors will emerge and capture the value.

PHYGITAL (PHYSICAL + DIGITAL) SCENARIO

The change we just discussed will not occur overnight but has already started. People will not change their perception of value in a short period of time but rather will start progressively perceiving the potential of digital assets while still valuing physical assets. A hybrid scenario will take place, the so-called 'phygital' scenario.

Phygital could describe anything that attempts to bridge the physical and digital worlds. It is mostly a marketing term to describe experiences. Consumers are themselves phygital. They exist in both the physical and digital worlds with multiple identities and participate in commerce in both places, often at the same time.

An interesting example of phygital is Amazon's expansion from being purely a digital player to becoming a phygital player by acquiring a bricks and mortar footprint. Starting from the digital world was the right play. Being digital by design means having an ability to scale quickly. Amazon's success is the best proof of the power of digital. However, most of the commerce volumes remain in the physical space and to fulfil the vision of being the "store for everything" Amazon had to be present in the physical space. Amazon acquired Wholefoods, a supermarket chain, to develop Amazon Go stores, a cash register free supermarket chain that recognises the majority of the value and transactions will still occur in an old-fashioned way over the short-to-medium term.

But other forms of value can emerge in the phygital space. That is where the digital twins comes to play.

WHAT IS
A DIGITAL TWIN?

HISTORY OF DIGITAL TWINS

IBM defines it as follow *"The digital twin is the virtual representation of a physical object or system across its life-cycle. It uses real-time data and other sources to enable learning, reasoning, and dynamically recalibrating for improved decision making."*[4] Generally speaking, it can be referred as a digital representation of a physical asset and its key attributes. It can be made of a digital document, audible or visual content or it could be the aggregation of someone's digital identity (digital footprint, professional or personal reputation, etc).

The concept emerged in the early 2000s with a publication from Michael Grieves of the Florida Institute of Technology. Grieves proposed to use digital twins as the conceptual model underlying product lifecycle management.

The digital twin concept comprises three different parts: the physical product, the digital product, and the connection between the two. They are connected by the data that flows from one to another. The information flow is fuelling the possibility for enhancing both the physical and the digital replica.

The concept has three different types: the digital twin prototype (DTP), the digital twin instance (DTI), and the digital twin aggregate (DTA):

- **Digital twin prototype.** The DTP represents the designing, analysing and processing leading to the materialisation of the physical product. The DTP is the very first iteration before the physical product.

- **Digital twin instance.** The DTI is the digital twin of every single instance of the created product.

[4] https://www.ibm.com/blogs/internet-of-things/iot-cheat-sheet-digital-twin/

- **Digital twin aggregate.** The DTA is the aggregation of DTIs whose data and information can be used for interrogation about the physical product.

The advantages of digital twins are obvious in heavy industries. Creating a digital twin of a factory and its production line can lead to finding ways to fine tune the production capacity while mitigating the risks of breakdowns. Equipping this factory with connected sensors (Internet of Things – IoT) can help identify vulnerabilities in the production lines as well as utilities and the operation of preventive maintenance. This significantly reduces the risk of having to stop production for remediation maintenance.

In that instance, the digital twin can be used as "patient 0" on which to run a series of tests to understand how the physical asset would react to changes. Digital twins are used in conjunction with historical data and human expertise to improve the outcome.

In a more service-oriented value chain, a digital twin could be a robotics process automation where a machine will perform the tasks of a human. To be successful, the process to automate must be well documented, have a well understood and measurable outcome and, most importantly, it must be replicable by a machine.

But let's take a step back and think about how we might create a digital twin to generate value in our personal lives. Let's take a look at what machines can replicate.

REPLICATION OF KEY ATTRIBUTES

It could be the replication of key attributes. For many of us key attributes are our physical attributes (age, gender, height, weight, education, skills, health, medical history, financial history, social class or groups we are associated with, etc). A digital twin can be the replica of these attributes to enhance our health and wellbeing. A digital twin could be a replica of our spending patterns in order to improve our financial wellbeing against predetermined goals.

REPLICATION OF IDENTITY

In the digital era, identity is a vast and complex topic. Proving that you are who you say you are to access to services is challenging. In the physical world we have solved this problem by enforcing proof of identity with public authorities' custodianship. At birth, we are being given a certificate and then a proof of identity in the form of an ID card or a passport. This grants us (or not) access to a wide range of services and products. But Identity goes far beyond the notion of proving who you are by showing an ID.

Identity is extremely complex and there are as many identities as there are human beings. If we add psychology on top, we become identity polymorph beings. We all have multiple identities, personal, professional, social, aspirational. We are quite different with our inner circle of trusted people (partner, family, friends) than we are with outer circles (colleagues, representatives of third parties, strangers, foreigners, etc).

Identity also morphs with perception and biases. Our professional identity is influenced by the industry we work in, our organisation, our past experiences and the codes that apply.

Ask yourself if your professional identity has changed since you were an intern or a graduate. The answer would probably be yes.

What you think your identity is might also differ from what your colleagues think your identity is.

Reputation is also a factor affecting one's identity. Reputation is the perceived trustworthiness of an individual. An individual with a reputation may be treated differently than his or her peers. A celebrity might be given priority over an average joe for a table at a fancy restaurant.

In the digital world, identity proves more challenging as it pushes the boundaries of our traditional understanding of this notion. But digital also enhances our ability to build and leverage our identity.

Digital identity is a vast complex world. Traditionally, public authorities had control over identity but in the digital world it is a different story. We all use a multitude of online profiles, email addresses, passwords, and social media accounts on which we display different traits and patterns.

Let us use the example of a fictional character called Jane Dow. To interact with public authorities, Jane uses a digitised version of her official identity, an email address with a password. In this instance, Jane's identity is defined by what is displayed on her ID, birth certificate, marriage certificate, tax file number, social security number, registered address, etc.

To connect with her professional network, Jane uses her LinkedIn profile where her identity is defined by her education, career path and the posts she publishes. On this platform she is a partner in a respected law firm and consistently posting thought leadership on the legal challenges associated with digital identity theft.

To connect with her personal social network, Jane uses Facebook, WhatsApp and Snapchat where she is known as JaneJane33 — a caring mother of two, a celebrity flash mob organiser in techno music festival, and a "petfluencer" in the cat wellbeing niche.

To interact with digital marketplaces for online shopping, Jane uses her Gmail account and mostly "logs in with Google". (Technology giants like Google, Facebook and Apple are emerging global players in the field of digital identity.)

This example illustrates that Jane Dow, like most of us is a polymorph individual and can be recognised by multiple identities based on different contexts or audiences.

This multitude of digital forms of identity presents challenges and immense opportunities for institutions to understand their customer base. From an interaction and personalisation perspective, the more data they collect the more they can profile and predict the needs of their customers.

It also presents a challenge from a compliance perspective. Most regulated industries have to run customer onboarding process known as KYC (Know your Customer). This enforced due diligence is designed to fight against fraud and forms of money laundering.

REPLICATION OF DIGITAL TRUST AND DIGITAL REPUTATION

Before diving into the power of digital reputation and how it is a key attribute/enabler for a digital twin, we'll explore the fundamental concept of trust.

"If money is the currency for exchange, trust is the currency for interaction" points out Rachel Botsman[5], academic, author and acclaimed public speaker. Rachel Botsman has focused her research on this cornerstone of human interaction that trust represents. Her book *Who can you Trust*, explores how the digital era is revolutionising human trust.

---

5 https://rachelbotsman.com/books/

Trust evolves, following societal development. It has traditionally been forged with one-on-one and face-to-face interactions. You earn trust as time passes. When societies were organised around small villages, hamlets or small cities, knowing each other and their trustworthiness was not a problem. Someone's reputation was a good indicator of whether someone could be trusted and was safe to interact with. As urbanisation and large cities started to form, knowing everyone and maintaining one-on-one interaction became impossible. We then invented "brands" as a proxy for trust.

A brand conveys a message about the product or service attributes, quality, consistency, price, etc.

We all know that our financial service providers are responsible people/entities because they are highly regulated and reputational. We can, therefore, trust them with keeping and protecting our money and assets. I trust that my insurance will remediate a damage to my property or car. I trust that my bank will not stop me if I want to withdraw my money.

This feeling is enhanced by a bricks and mortar presence. If something goes wrong, you can always go to the closest branch and take your money out.

The digital world challenges the traditional construct of trust. Sure, a well-known brand can go digital and be trusted but the digital world is the Wild West. How can we be certain that it's not a scam or the individual or entity we want to interact with is trustworthy? This concern is even more prominent in the gig economy. If one needs to hire a freelancer, how can one know if that person is qualified, will do a good job and charge the quoted amount? If one needs to hop on a ride share, how can one know that the ride will be safe? If one's monetising a spare bedroom, how can one know if the guests will be respectful and will not damage the place?

Platform businesses like Amazon, Uber, Airbnb, etc have popularised the "star rating" mechanism to enable users to rate the quality of their interaction. This has become a component of digital reputation and a proxy for trust. After all, if an individual is well rated and, therefore, trusted by people with whom they interacted with for a short period of time in the digital economy, then the star rating should reflect trustworthiness.

In financial services trustworthiness can be associated with level of risk and, therefore, pricing. In 2018, a fascinating experiment was run by a bancassurance company. The experiment consisted of trying to understand if someone's digital reputation (the aggregation of the feedback, star rating received in the digital economy) was a good indicator of risk. In order to do that, this incumbent partnered with a start-up that specialised in digital reputation, Traity.

Traity had built an algorithm that worked with a user's permission. Should the user grant access to a broad range of digital profiles (Airbnb, Uber, LinkedIn, etc), the algorithm would verify the identity and aggregate the scores received. That algorithm had been trained with four million participants.

The incumbent was willing to test if that type of data could be ingested and if yes, would it enhance its pricing algorithm.

The results were fascinating, digital reputation had enabled the improvement of the pricing algorithm on certain perils. Meaning that, digital reputation could be used to better understand and price risk.

This experiment proved the value of this emerging form of trust and its potential application in better understanding risk.

In some of the ethnographic research done on the concept of digital reputation some findings indicated that people were valuing their ride sharing score and willing to protect it and adapt behaviour when interacting to mitigate the risk of an unfavourable review. Anecdotally, a young lady stated that when she had had too many drinks, she'd rather pay more and take a cab than hop in an Uber and get a bad review.

We can then assume that digital reputation is an emerging form of trust and, as we'll see that at a later stage as an emerging asset class.

Start-ups like Traity[6] and Reputationaire[7] (to which I have the privilege to serve as a Board Advisor), have specialised in helping people leverage their online footprint and digital reputation to get better access to services across such as finance, insurance, recruitment and rental property.

You can leverage digital reputation, which is a form of digital twin, to empower millions of people to (poorly served, underserved or unserved) gain access to basic services that will change their lives.

REPLICATION OF KNOWLEDGE

Replication of knowledge is probably one of the most interesting key personal attribute that will constitute a digital twin.

What is more personal than knowledge? It is the result of an entire life of education, experiences and exploration. It defines parts of who we are as individuals and professional. Replicating that knowledge is the most obvious and valuable aspect of digital twins.

It is also probably the most complex and ambiguous part. There is no clear definition of "knowledge" because it is a by-product

---

6 https://www.crunchbase.com/organization/traitperception

7 https://reputationaire.com/

of so many variables (education, personal values, environment we evolve in, personal interests, personal connection, fields of expertise, etc).

From a perspective of monetising a digital twin, one of the most obvious way to start is to digitise professional expertise and empower the digital twin with it.

Let's say that you operate in a field of expertise that is heavily defined by regulations, laws or customs but leaves parts open to interpretation. Expertise in accounting, auditing, law practices, for example, are particularly prone to being digitised and scaled because of their well bounded nature. Your personal experience in your career brings depth and a unique touch to your digital twin.

Current technology enables us to train a machine to replicate components of a knowledge in about two weeks. This is the investment of time you should spend to "download" your expertise to your digital self.

Replicating your academic knowledge is probably a waste of time as it's easy to foresee university course modules as downloadable and consumable by your digital twins. The application and testing of the theory in reality is what makes you unique.

REPLICATION OF BEHAVIOUR AND THOUGHT PROCESSES AND BIASES

In a similar fashion to the replication of knowledge, our psychological profile and biases will be replicated by our digital twins. To be trained, during the replication process, your digital twin will ask a series of questions designed to map out your brain connections as well as your thought process.

Again, this is one of the layers of yourself which makes you unique and will give value to your digital twin.

REPLICATION OF INTELLECTUAL PROPERTY

In this chapter we have only scratched the surface of some of an individual's attributes that can be replicated and transferred to a digital twin.

All we have just described can be assimilated as your intellectual property (IP). In order to leverage the power of digital twins, user groups (you, me, any person interested in monetising our knowledge) will need to find a way to "licence" our IP to our digital twins. It is quite unlikely that digital twins will be available in open source but rather being offered as a commercial service by a specialised technology company.

It is plausible that some law firms will add "digital twin IP licencing" to their existing suite of services. It may coexist with more well-known offering such as commercial litigation, divorce, etc.

Without a proper way to protect our IP we'll simply give away our deepest identity to technology companies and we will definitely be replaced by machines.

In this chapter we have explored what a digital twin will be. This exercise has been theoretical. In the next chapter we explore real life examples of how digital twins are being used.

WHERE ARE DIGITAL TWINS CURRENTLY USED?

Digital twins were born in complex environments to augment or supplement the human brain with a specific purpose to protect expensive physical assets.

Mostly digital twins are used to replicate attributes of physical assets. They are used in mining, manufacturing, facility management and heavy machinery, for example. These digital replicas are powered by series of sensors that feedback vital information to a central command such as:

- Heat sensors ensure a machine operates within a manufacturer's recommended range. Below or over the recommend threshold, the likelihood of a fault, defects or a breakdown increases.
- Speed sensors ensure will ensure that the engine is working to the safest and best possible range of reps. It will maximise productivity and minimise risk of failure.
- Vibration sensors feedback any anomalies that could indicate early signs of potential breakdowns immediately.

Digital twins have proven their efficiencies in preventive maintenance. This leads us to a fundamental concept upon which some business models are built — remediation and prevention. These business models fuel trillions of dollar industries like the:

- **Health industry.** The health industry spans hospitals, general practices and treatments for existing physical and mental health conditions.
- **Pharmaceutical industry.** The pharmaceutical industry is responsive by nature. It develops drugs, vaccines and supplements to remediate known diseases. These companies are in the business of healing people through applied chemistry.
- **Insurance and reinsurance industries.** Insurance and reinsurance are very similar in their business model. These industries exist to remediate damages and compensate loss for physical assets.

An obvious way for these industries to leverage the power of digital twins is to enable them to expand or shift their business models from being purely remediation centric to active players in the space of prevention.

One growth pathway is using digital twins to create digital repositories or replicas of physical assets or persons.

A digital health record topped with a genome test will enable any person and their general practitioner to prevent disease and extend life. On that topic I highly recommend *Lifespan, the revolutionary science of why we age and why we don't have to* by David A. Sinclair, PhD.

In his ground-breaking book, David Sinclair, a Harvard Medical School researcher explains how ageing is a disease like any other and can be treated to expand life. A genome test, that nowadays costs few hundred dollars, can help predict the most likely health threat and develop tailored treatments.

A digital replica of your health history and genome profile can be used to not only heal you but also help science learn more. The fundamental ethical questions (around data privacy) raised by such a model are explored in a later section.

Another example of where digital twins can power new business models is in the insurance industry. This industry exists to remediate damages on physical assets. For insurers accessing a digital replica of a car's maintenance history or a property construction material and maintenance record can be crucial to pricing risk.

It can also be crucial to understand the likelihood of failure and help prevent it for the benefit of the asset owner. Insurers are perceived to be in the business of collecting premiums and trying not to pay claims, but they are not. It is in the best interest of the stakeholder they serve (customers, shareholders, employees, community) to avoid claims. Less claims means less damages, less accidents, less injured people, less lives broken or lost.

Insurers and reinsurers act as a shock absorber for an economy. Moving to prevention will help them provide more meaningful services to their customers.

Accessing, orchestrating or maintaining an inventory of their customers' digital twins (applied to physical assets) will let an insurer know that a service is due on their car and preventive maintenance is due on a property.

They will be in a position to incentive the customer to have their car maintained. They could even offer maintenance subscription services for properties and cars.

All the collected information will feed back to the digital twins and the data lakes fuelling pricing and underwriting capabilities.

In order to position themselves for success under this new business model, insurers are currently taking stakes into telematic capabilities as well as connected environments emerging players. These are early forms of digital twins.

Telematics are onboard sensors in a vehicle that tracks key indicators such as speed, brakes (how often used and the intensity), routes, deviations, etc. It enables insurers to identify driving patterns and fine tune their price. It also enables them to understand how the asset is being used and create predictive models for breakdowns.

Connected environments describe places (individual or commercial properties) equipped with sensors monitoring temperature, airflow, utility usage (water, gas, electricity). By monitoring these parameters, preventive actions can be taken to reduce or eliminate risks.

A great example is water leaks. Water leaks often go undetected for extended periods of time (months, year, decades). Not only does it waste the most precious resource on Earth but it can cause extensive damage to building structures over time. Sometimes it renders a place unsafe. A pipe bursting can also create a flash flood and expensive damage to a property.

Connected devices like Flo[8] are placed on a main water pipe to identify microleaks. If detected, it will proactively shut down the water flow, send a notification through the App and connect a customer with a local plumber.

Not only it would enable insurers to create an ecosystem of service for their customers but it also prevents further damages on the property. Water related damages cost billions of dollars to insurers every year and this could be prevented with similar sorts of devices.

Mould and mildew is another example of how risk can be prevented using early forms of digital twins.

Mould is one of the most common, widespread organisms on the planet. It is everywhere. On the ground, on your skin, in your body, in the air. It acts as the planet's digestive system, feeding itself on rotting organisms and recycling them. Without mould we wouldn't have products like cheese, beer, wine, etc.

Around 50,000 species of mould exist but only seven are harmful or deadly to humans. It is a significant health hazard if a harmful mould

---

8 https://meetflo.com/

species grows in a property. It can cause people with respiratory conditions like asthma to be more prone than others to suffer more asthma attacks. Mould grows everywhere and is inevitable. In order to identify if a species is harmful, it requires an expensive lab test. Although harmful mould is unlikely, the risk remains of it growing if you don't watch out for it.

Four factors are required for mould to grow:

- A source of food (which could be anything or any type of material).
- A lack of direct sunlight.
- A humidity level over 70%.
- Poor ventilation.

That is the reason why mould usually thrives in bathrooms or basements.

If one of the four parameters change then mould cannot grow and the risk disappears.

If a sensor in an air ventilation or airconditioning system senses the humidity level at above 70%, it can automatically start the airconditioner or ventilation system so it drop to 69%. This drops the risk of mould to almost zero.

This is another great application of early forms of digital twins solving real life problems.

Endless applications can be developed for digital twins in property management and this can lead to developing safer places, and a more sustainable and intelligent environment in buildings.

DIGITISING PHYSICAL ASSETS

Another real-life example[9] of how digital twins can deliver immense value is in digitising physical assets and facilitating or automating ecosystems of services.

Insurance is a very conservative industry which is struggling to digitise its value chain. The industry is built around products to protect physical assets. Customers who own these assets are treated as attributes of the insurance products and are not the centrepiece of the product end-to-end lifecycle.

When a car insurance policy holder experiences an insurable event (accident, theft, weather event related damage) and lodges a claim, it triggers a complex process. This process depends on the information flow needed by the insurer to process the claim. In short, the insurer will ask questions about the asset first and then about the customer. It will lead the insurer to investigate the claim, engage professional assessors to gather evidence to make a decision to accept or reject a claim. The customer will be asked to pay the excess first in order to trigger the claim process.

The insurer will estimate the damage by running an assessing sub process. This involves either inhouse resources or external vendors. The assessor will go on site to assess the damage to the vehicle. They use third party software to estimate the damage to body and mechanical parts. The assessor then submits a report making recommendations and with a cost estimate.

In the meantime, the insurer will try to establish liability to understand if the customer is at fault or if the other party was. Fraud checks are held in the background.

---

9 https://www.ledgerinsights.com/state-farm-usaa-auto-insurance-claims-blockchain/

In most instances another or multiple parties are involved in an accident. This increases the number of insurers involved in the process. Determining liability becomes more complex. Interviews are held and police reports gathered. If some form of injury occurs, medical reports, ambulance and hospital bills are collected.

Insurers of the parties involved run this process in isolation and eventually call each other to share their conclusion and estimate individual risk positions.

Another round of communication with the customer keeps them informed about the progress. Eventually the claim will be authorised or denied, and repairs conducted following the customer's decision.

Some insurers own repair centres to keep costs under control. They will order parts and do the repairs. In other scenario, the repair will be conducted by a third party repairer who invoices the insurer.

Based on agreed liability the insurer will then recover the cost with the third-party insurers. This process, which has been bastardised on purpose, takes anywhere from between few weeks to few months. Between 2016 and 2018, in Australia, four major insurers aggregated +80% of the market, so this can take up +90 days. And during this, the customer is left without a vehicle.

The most incredible part of this process is that almost nothing is automated. Insurers spend hours on the phone identifying the insurer of the other party involved in an accident and who is liable. They exchange countless emails and written documents before agreeing on a final position.

Digital twins can automate that process.

In fact an experiment run out of Australia, involving the four major insurers and Everledger[10] (a tech company specialising in using distributed ledger technology (DLT)) demonstrated using blockchain to create digital twins from existing assets. A strength of DLT is it is an unalterable record of an event. It creates an unalterable single source of truth. Blockchain also permits the development and use of automated contracts. By applying this technology to the claim process can reduce the average handling time of a claim from 90+ days to 2 - 3 days.

Given their highly regulated environment, insurers have always had difficulty securely disclosing and sharing data to enhance interaction between their customers. Insurers hold a lot of data about 'objects' (properties, vehicles) and about their lifecycle. They use this data for individual competitive advantage, to price risk and process claims.

There are times where data needs to be legitimately shared. Traditionally it's been done using a manual pull process. But this process is cumbersome and can lead to improperly disclosing data.

The experiment consisted in building a private blockchain (we'll explore blockchain as an enabler for digital twins in a later part of this book) containing all insured vehicle key attributes (VIN, make, model, age, insurer, information about the owner, etc). The Regulator could only administrate this secure database and only insurers with a legitimate reason to interrogate the database would receive authorisation. Insurers could then share encrypted and anonymised legitimate and trustworthy data. This central depository of information would also be used to connect insurers involved in a claim. The answer provided would enable the trigger of the next step of a process, even automating claims. Because of the use of blockchain, all interaction

10 https://www.everledger.io/

would be captured, stored and protected. This would render all interactions easily auditable for a regulatory body which could ensure appropriate use of data.

After months of experimenting, and two independent third parties testing and validating the technology, they identified 27 different values pools could be unlocked (fraud, subrogation, vehicle lifecycle and provenance, asset financing, etc). At scale, this would enable insurers to better serve customers and save hundreds of millions of dollars annually across the value chain. It was also identified that it could create opportunities for insurers to expand their traditional remediation business model to prevention and safety. Hence bringing more value to their customers and a right to play in having multiple meaningful interactions a year. Under the current remediation business model, an insurer only has one opportunity (if no claims occur) to interact with their customers. Any interactions with an insurer are unpleasant. It is either to renew a payment or to deal with the consequences of a terrible event during a claims process.

Two major American insurers (USAA and State Farm[11]), announced early 2021, that they were operationalising this concept. They are aiming at automating 79,000 annual claims using blockchain technology as an enabler.

In this chapter we have seen examples of where digital twins are already in use. In the next chapter we'll explore how can they be scaled, and which technologies will enable them.

[11] https://www.ledgerinsights.com/state-farm-usaa-auto-insurance-claims-blockchain/

WHAT TECHNOLOGY DO DIGITAL TWINS RELY ON?

As we have already explored, digital twins, in their current state, mostly rely on IoT sensors that replicate key attributes of physical assets.

In their current form, digital twins do not focus yet on replicating humans but rather on what they own and value.

This is changing fast, enabled by technologies like big data, artificial intelligence (AI), natural language processing (NLP), machine learning (ML), augmented reality (AR), virtual reality (VR), holographic technology and digital humans.

These enablers are only the ones we know of today but technology that does not yet exist or, at least, is not at scale will accelerate the development of digital twins. Technology like quantum computing.

This chapter discusses these technologies and highlights their enabling potential for digital twins to thrive. Like a human body, digital twin will use fuel for energy, brain for decision making, a conscious to continue learning and develop, they will have a visual appearance to interact with other forms of beings and they will have a memory.

DATA WILL FUEL OUR DIGITAL TWINS

Big data is closely associated with storage technology. The past 20 years have seen the rise of digital storage from flash drives, portable hard disks, computer servers and mainframes, digital tapes, DVD/Blu-ray to where we now have cloud-based data centres with unlimited storage capacity.

The term "big data" is credited to John Mashey[12], a US computer scientist who coined the term in the 1990s. Big data is a term used to describe datasets with sizes beyond the usual ability of software to triage, curate, process and extract intelligence within an acceptable time. The notion of big data is then relative to the user's technical ability and time availability. The philosophy of big data encompasses unstructured, semi-structured and structured data; however, the main focus is on unstructured data. Size varies from terabytes to zettabytes of data. According to a 2018 definition, 'big data is where parallel computing tools are needed to handle data'.

Big data has increased the demand for information management systems that enable the collection, processing, analysis and visualisation of data to make it intelligible as a basis for making business decisions.

Developed economies are more and more data intensive. Billions of mobile phones are capturing users' behaviour, movement, interests and web searches. While smartphones became a proxy for understanding a user's psyche, IoT devices like smartwatches have become a proxy for tracking and understanding their physiology. For example, heart rate, blood pressure, activity, sleep time, quality of sleep, level of activity/inactivity.

Let's now try to use numbers to understand the magnitude of data flooding the world's economy:

- Between 2018 and 2020, 90% of the world's data was created. Now 5G technology will increase that trend.
- Google processes over 3.5 billion search queries every day[13].

---

12 https://en.wikipedia.org/wiki/John_Mashey

13 https://www.internetlivestats.com/google-search-statistics/

- It was forecast that by the end of 2020, 44 zettabytes will make up the entire digital universe (44 000 000 000 000 000 000 000 bytes).
- There are 2.5 quintillion bytes (2 500 000 000 000 000 000 bytes[14]) of data produced every day.[15]
- Every day, 306 billion emails and 500 million Tweets are made[16].
- By 2025, each human will produce 463 exabytes of data each day.

All that data can be contextualised, analysed and acted upon. It represents endless possibilities for understanding people — who they are, who they think they are, what they know, what they think they know, what they don't know, how they think.

Fundamentally the ability to mine, analyse and contextualise large amounts of structured and unstructured data will fuel digital twins. They will be our digital replicas that can be augmented with all the world's available knowledge.

ARTIFICIAL INTELLIGENCE WILL OPERATE AS THE BRAIN OF OUR DIGITAL TWINS

AI is a broad term that associated a form of problem solving achieved by non-cognitive intelligence. The Oxford Dictionary defines AI as 'the theory and development of computer systems able to perform tasks normally requiring human intelligence, such as visual perception, speech recognition, decision-making, and translation between languages. This intelligence can be utilised to replicate tasks with a well-established measurable and repeatable outcome.'

14 https://www.raconteur.net/infographics/a-day-in-data/

15 https://www.socialmediatoday.com/news/how-much-data-is-generated-every-minute-infographic-1/525692/

16 https://www.internetlivestats.com/one-second/#email-band

AI is not a new concept. It has been in collective consciousness since antiquity. In Greek mythology, the Island of Crete was were Talos, a bronze made automaton was patrolling the coastline in order to protect Europa from invaders. In more recent times, a form of AI is depicted in the form of a semi living being in Mary Shelley's book Frankenstein, or the Modern Prometheus published in 1818. In the 1950s, Isaac Asimov (famous science fiction writer and professor of biochemistry at Boston University) wrote about a world where machines are not only intelligent but also flirt with consciousness.

The story of Frankenstein, like many similar stories raise some issues currently discussed in AI ethics. These think tanks are trying to avoid the risk of machine led discrimination and built-in bias.

Contemporary to Asimov's writing, AI was established as a field of research in 1956 at Dartmouth College, where the term "artificial intelligence" was coined by John McCarthy[17] to distinguish the field from cybernetics. John McCarthy along with Allen Newel[18], Herbert Simon[19], Marvin Minsky[20] and Arthur Samuel[21], became the founders and leaders of AI research. With their students, in 1954, they produced computers that were learning checkers strategies. It was reported that by 1959, the computers were playing better than the average human. In 1956, their program had learnt some basic theorems and were solving algebra problems. In the 1960s, their research was accelerated using funding from the Department of Defence. During that time Herbert Simon predicted that '*machines will be capable, within twenty years, of doing any work a man can do*'. Marvin Minsky agreed, writing, '*within a generation ... the problem of creating 'artificial*

17 https://en.wikipedia.org/wiki/John_McCarthy_(computer_scientist)

18 https://en.wikipedia.org/wiki/Allen_Newell

19 https://en.wikipedia.org/wiki/Herbert_A._Simon

20 https://en.wikipedia.org/wiki/Marvin_Minsky

21 https://en.wikipedia.org/wiki/Arthur_Samuel

intelligence' will substantially be solved'. In 2021, we can agree they were wrong from a timing perspective, but we can't disagree with their vision which is materialising every day.

ARTIFICIAL INTELLIGENCE DEVELOPMENT TIMELINE

In the 1970s, government funding (in the US and UK) did shrink due to slow progress in the field of AI. This field was revived in the 1980s in light of the commercial success of "expert systems", which was a type of AI program that simulated the knowledge and analytical skills of human experts. In the late 1980s, progress in transistor technology enabled the development of the artificial neuronal network (ANN) which boosted AI research.

In the late 1990s, AI began to be used for logistics, data mining, medical diagnosis and other areas. The success was due to increasing computational power, greater emphasis on solving specific problems, new ties between AI and other fields (such as statistics, economics and mathematics), and a commitment by researchers to mathematical methods and scientific standards. Deep Blue became the first computer chess-playing system to beat a reigning world chess champion, Garry Kasparov[22], on 11 May 1997.

In 2011, IBM's Watson defeated the two greatest Jeopardy! champions.

AI entered a new era in 2015 when AlphaGo won four out of five games of Go in a match with a Go champion. AlphaGo is a program that plays the boardgame Go, which is far more complex than chess as it presents far more alternatives to consider per move. (AlphaGo was developed by DeepMind Technologies and later acquired by Google.) This was a significant milestone as AlphaGo became the first computer Go-playing system to beat a professional Go player.

---

22 https://en.wikipedia.org/wiki/Garry_Kasparov

By 2017, a survey highlighted that 1 in 5 companies reported they had incorporated AI in some offerings or processes.

ARTIFICIAL INTELLIGENCE REPLICATES PREDICTABLE OUTCOMES

Countries are now making huge investments in the field to develop AI dominance. Some reports are that China invested $70bn in AI research in 2020. In 2017, China released a three-step program aimed at positioning the country as an AI world leader by 2030.

AI is used in healthcare to predict strokes, in banking and insurance to interact with customers through chat bots, for process automation (including claims), to create risk predictive models, in social media to predict relevant content and suggest connection, in advertising to target audience, etc.

At its fundamental, AI replicates predictable outcomes that do not require human cognitive abilities (we'll see in next sections how ML and NLP changes the dynamics).

AI will be a core technology upon which digital twins will thrive. They will be capable of understanding our traits, patterns, assets and consumption to replicate decisions for us.

AI is so powerful that a not-for-profit organisation called OpenAI[23] has committed, through its purpose, to ensure that artificial general intelligence (highly autonomous systems that outperform humans at most tasks) will benefit all of humanity and not just a few. OpenAI is funded by investors such as Microsoft, Reid Hoffman's charitable foundation and Khosla Ventures.

---

23 https://openai.com/about/

OpenAI has developed "GPT-3" (generative pre-trained transformer 3), an AI trained with 175 billion parameters. To put this into perspective, the previous release (GPT-2) was trained with 1.5 billion parameters and was considered state of the art only a year before. That represents an 11,566% increase in parameters used for training GPT-3. It is capable of writing creative content close to a human level. The machine demonstrated ability to handle abstraction. GPT-3 can also handle conversation with humans that feels natural.

GPT-3 can also, in a few seconds, generate code from a high-level description the developer gives it. The human developer just needs to ensure the relevance of the written code and copy paste it into the program in development.

GPT-3 has also demonstrated capability for handling business. The machine was asked to create a paper on how to recruit Board Members and provided a three-step process.

In order to perform at such as amazing level, GPT-3 has absorbed all the "text" available on the internet. Everything that has ever been created and is publicly available has been used to train that machine. That unconceivable amount of information (at least for a human brain) has been ingested, stored and can now be contextualised and applied to pretty much any scenario that we would throw at the machine.

The GPT-3 is, in some ways, the ultimate digital twin of the human collective psyche.

What could be perceived as the early stage of "Skynet" (the evil corporation that built the Terminator in the eponym movie from James Cameron) isn't (yet) quite a threat to humanity. Although GPT-3 can demonstrate some level of abstract thinking, it still isn't capable of deep expertise in that space. Putting it bluntly, GPT-3 still lacks common-sense. However, it isn't a gamble to predict that gap will be bridged in the very near future.

We should make no mistake; AI is going to change the world and GPT-3 is just a glimpse of what is coming.

MACHINE LEARNING AND DEEP LEARNING WILL BECOME THE CONSCIOUS OF OUR DIGITAL TWINS

MACHINE LEARNING

ML is used to describe an algorithm built to learn from its own mistake. It is a subset of AI. See AI and its subsets in Figure 1.

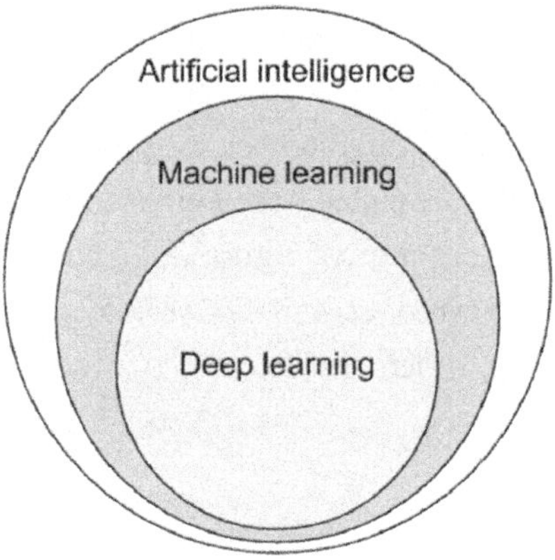

Figure 1 AI and its subsets

ML is a model based on sample data called "training data". This sample helps the algorithm make predictions or decisions without being directed to do so. Once the algorithm has been trained, it is released into production and will use larger database and continue learning.

ML is used in a broad range of applications such as email filtering, computer vision, automotive industry, financial services, etc. The power of ML surpasses human cognition. As an example, ML is used to train computer vision in automated vehicle technology. It has been trained to recognise driving related risks, such as objects or animals. In so doing the machine has been "shown" millions of pictures of animals from different species. It has been trained to identify their shapes, size, mass, colours, etc. The machine is now faster at recognising a domestic cat, in daylight, than a human.

A related field is data mining, which consists of exploring data analysis through unsupervised learning. This is also a key enabler for digital twins as they will have to exploit large volumes of structured, partially structured and unstructured data.

Machine Learning development timeline:

Logically AI preceded ML as a field of research. While the exploration of AI started in 1956, the term ML was coined in 1959 by Arthur Samuel, an American pioneer in both fields of AI and ML.

ML started with pattern recognition and classification. This trend continued until the 1970s. From pattern recognition the field of studies expanded to strategies. In 1981, there was a report on using teaching strategies so that a neural network can learn to recognise 40 characters from a computer terminal.

In 1997, Tom Mitchel[24], an American scientist and professor at Carnegie Mellon University, proposed the following quote to define ML. 'A computer program is said to learn from experience E with respect to some class of tasks T and performance measure P if its performance at tasks in T, as measured by P, improves with experience E.

[24] https://en.wikipedia.org/wiki/Tom_M._Mitchell

This definition is heavily oriented toward an operational outcome rather than a cognitive one.

This follows a paper *Computing Machinery and Intelligence*, published 47 years earlier, by Alan Turing[25]. He is the English mathematician that deciphered the German ciphering machine Enigma to access to critical intelligence about the war against the Nazis. Turing's story was brought to the screen in the movie Imitation Game released in 2014. In his paper, Turing reframes a paramount question from 'Can machines think?' to 'Can machines do what we (as thinking entities) can do?'

Modern day ML has two objectives. One is to classify data based on models that have been developed and the other is to make predictions about future outcomes based on these models.

It is important to acknowledge that at the time of writing, and probably for some time to come, not everything can be learned by a machine. Indeed, machines are trained by the most complex and imperfect being in the known universe — us humans. We are biased and don't possess universal knowledge or wisdom. The algorithm will be as good as its programmer and all accessible knowledge at a point in time. ML is also built to deliver a variety of tasks that can be automated. Here again, not everything can be automated. Decisions in context requiring human interaction and an emotional, cognitive answer cannot be automated yet.

It is also likely that human cognition will not be advanced enough to understand the answer of an ML algorithm. To solve a problem we have two options, ask a human or a machine. The machine will compute, analyse, organise and solve a problem that could take decades for a human to achieve. It is quite possible the conclusion drawn by the machine may not be intelligible for a human consciousness at that point in time.

---

25 https://en.wikipedia.org/wiki/Alan_Turing

A more likely, foreseeable scenario is a cohabitation between a human and its dedicated digital twin built-in ML. The more we spend some time training our digital twin to understand us, the faster it will start enhancing our lives. We will be able to choose to dedicate them to certain tasks and we'll perform others we deem more strategic, valuable and satisfying, etc.

Again, if this sounds like science fiction, consider that smart assistants like Google Assistant, Google Studio, Siri from Apple and Alexa from Amazon, are early forms of digital twins that are accessible to the mass market. They learn to recognise our voices; learn from our web searches and the requests we make so that they can become better at predicting and understanding who we are or who we aspire to be.

DEEP LEARNING

DL is a subset of AI. It's a specific field of research within ML. It can be supervised, semi supervised or unsupervised. ML is a learning method based on ANN. And ANN replicates the known distributed communication nodes in biological systems.

Applications for DL include medical image analysis, machine vision, speech recognition, NLP and audio recognition. In some fields, the results demonstrated by DL have sometimes surpassed human expertise.

Deep Learning development timeline:

Ukrainian mathematician Alexey Ivakhnenko[26] published the first learning algorithm for a supervised, deep, feedforward, multilayer perceptron (an algorithm for supervised learning of binary classifiers) in 1967.

---

26 https://en.wikipedia.org/wiki/Alexey_Ivakhnenko

Rina Dechter[27], a Computer Science Professor at University of California Irvine, coined the term "deep learning" in 1986.

In 1989, Yann LeCun[28], a French computer scientist, professor at NYU and currently Chief AI scientist at Facebook, experimented with a DL algorithm for hand-writing recognition. It successfully recognised handwritten ZIP codes on mail. The training of the algorithm took three days.

Two years later, in 1991 similar systems were used to recognise isolated 2D hand-written digits and recognise 3D objects. At the time, both development and the performance were limited by computation power and our knowledge of biological neuronal networks. An acceleration occurred in the coming decade with speech recognition being heavily funded by DARPA (the R&D arm of the US Department of Defence [DoD]) and the National Security Agency.

The impact of DL in industry began in the early 2000s. At the time it was estimated that DL based algorithms were already processing 10 to 20% of the handwritten cheques in the US (according to Yann LeCun). Large scale industrial applications of DL such as speech recognition started around 2010.

Advances in hardware drove renewed interest in DL. In 2009, Nvidia was involved in what was called the "big bang" of DL, as DL neural networks were trained with Nvidia graphics processing units (GPUs). That year, it was estimated that GPUs could improve the speed of DL systems 100-fold.

---

27 https://en.wikipedia.org/wiki/Rina_Dechter

28 https://en.wikipedia.org/wiki/Yann_LeCun

In 2014, Josef Hochreiter[29], from the University of Linz, used DL to detect off-target and toxic effects of environmental chemicals in nutrients, household products and drugs and won the Tox21 Data Challenge.

In 2015, Google's speech recognition reportedly experienced a dramatic performance jump of 49% through connectionist temporal classification trained long short-term memory, which they made available through Google Voice Search.

In March 2019, Yoshua Bengio[30] (University of Montreal), Geoffrey Hinton (University of Toronto) and Yann LeCun (New York University) received the Turing Award for conceptual and engineering breakthroughs in their work in recognising deep neural networks are a critical component of computing.

Since then, DL is used in:

- Medical image analysis to identify cancer tumours.
- Drug discovery and toxicology. In 2019, it was used to produce molecules that were validated experimentally using mice as test subjects.
- Customer relationship management, deep reinforcement learning has been used to approximate the value of possible direct marketing actions.
- The automotive industry, it has been used to identify threats such as animals on the road.
- In financial services for fraud detection and anti-money laundering.
- The US DoD applied DL to training robots to perform new tasks through human observation.

---

[29] https://en.wikipedia.org/wiki/Sepp_Hochreiter

[30] https://en.wikipedia.org/wiki/Yoshua_Bengio

DL will be a cornerstone for developing our digital twins. After you train them to replicate your traits and thought processes, they will improve by learning from our outputs and their own. They will be able to speak any language to interact with humans. They will also "speak" digital languages to interact with machines. They will have the capacity to translate, contextualise and make sense of things for us when we interact with other machines (digital humans).

Now that we have explored the foundation technologies that will power the cognition abilities of our digital twins, lets now dive into technologies that will enable visual interactions.

AUGMENTED REALITY

AR is a way to experience the environment, where objects are enhanced by computer enabled information. AR can be multi sensorial, for example, visual, auditory, olfactory. AR is also a combination the real and virtual worlds in real time and in 3D. The use of AR alters the way we experience the real world.

The primary value of AR is the manner in which components of the digital world blend into a person's perception of the real world, not as a simple display of data, but through the integration of immersive sensations, which you perceive as natural parts of an environment.

One of the best and most common applications is Google Lens which, while using the camera of your smartphone, can:

- Recognise an object you are pointing at.
- Give you information about that object such as its attributes (make, brand, material, price, where to shop).

One of the most popular commercial applications of AR is Pokémon Go. The game became a worldwide phenomenon in record time in the mid-2010s. To augment your ranking and maximise chances to defeat competitors, a player needs to catch as many Pokémon as possible. Each have specific attributes, powers and weaknesses. Pokémon Go was a mobile game, in which, players had to use their smartphone lens to track Pokémons in specific locations. Suddenly, cities around the world experienced flash mobs in random locations with people were searching for Pokémons using their phone. There were incredible stories reported in the news. People stopping on highways to hunt Pokémons. Players creating traffic accidents or traffic jams. Some players even left their cars unattended in traffic to catch Pokémons in rare Pokémon hideouts.

The US Air Force developed the earliest functional AR systems to provide an immersive mixed reality experience[31] for users in the early 1990s. AR was used to overlay sensory information in workplaces. In this context we talk about pilot training. The intent was to test how human performance could be enhanced when operating various tasks and bring forward information to reduce the mental workload.

Originally AR commercial applications were in the entertainment and gaming industry, then it became widely used in education, communications and medicine.

The applications of AR are endless. For example, imagine a surgeon in the operating theatre focusing on the operation with the patient's vital signs projected on to their peripheral vision. Imagine going grocery shopping and, while walking on the aisles, using a filter that only searches for food that matches a specific diet. The hardware and software necessary for these scenarios already exists.

---

[31] Armstrong Laboratory https://apps.dtic.mil/dtic/tr/fulltext/u2/a292450.pdf

THE VISUAL INTERFACE OF OUR DIGITAL TWINS

VIRTUAL REALITY

If AR overlays a virtual filter over a real environment, VR provides the live experience in a virtual one.

Like AR, VR applications are useful across multiple industries, including medicine, entertainment, gaming, military, education, aeronautics and mining.

Using VR to simulate reality it can train people to safely operate expensive assets such as planes, vehicles and heavy machinery. The technology replicates the reactions and motions of the equipment someone is operating in the simulated environment.

Avatars are also making virtual social gathering and interactions possible. Facebook is currently working on creating virtual spaces (room, theatres, restaurants, concert arenas) where users can meet a friend's avatar to interact.

Hardware enabling VR experiences vary from a simple display (smartphone or laptop) or a head mounted device (HMD) such as an Oculus Rift headset, to entire rooms where visual and sensory experiences can be created (plane cockpit, control centre room, etc). The first HMD was created by Ivan Sutherland[32] from University of Utah in 1968. The device was so heavy that it had to be suspended from the ceiling.

---

32 https://en.wikipedia.org/wiki/Ivan_Sutherland

Virtual Reality development timeline:

The VR industry mainly provided devices for medical, flight simulation, automobile industry design and military training purposes from 1970 to 1990.

By the 1980s, the term "virtual reality" was popularised by Jaron Lanier[33], an American computer scientist, computer philosophy writer and visual artist. He was one of the modern pioneers of the field. Lanier founded the company VPL Research in 1985. VPL Research developed several VR devices such as the DataGlove, EyePhone and the AudioSphere.

The 1990s saw the first widespread commercial releases of consumer headsets. In 1992, for instance, *Computer Gaming World* predicted "affordable VR by 1994".

In 1991, Sega announced the Sega VR headset for arcade games and the Mega Drive console. It used LCD screens in the visor, stereo headphones, and inertial sensors that allowed the system to track and react to the movements of the user's head.

The same year Antonio Medina, a MIT graduate and NASA scientist, designed a virtual reality system to "drive" Mars rovers from Earth in apparent real time despite the substantial delay of Mars-Earth-Mars signals.

In 1992, Nicole Stenger created Angels. The first real-time interactive immersive movie where the interaction was facilitated with a dataglove and high-resolution goggles. That same year, Louis Rosenberg[34] created the virtual fixtures system at the US Air Force Armstrong

33 http://www.jaronlanier.com/

34 https://en.wikipedia.org/wiki/Louis_B._Rosenberg

Labs using a full upper-body exoskeleton, which enabled a physically realistic mixed reality in 3D. The system enabled the overlay of physically real 3D virtual objects registered with a user's direct view of the real world. This produced the first true augmented reality experience that enabled sight, sound and touch.

Closer to today, in 2010, Palmer Luckey[35] (an American entrepreneur) designed the first prototype of the Oculus Rift. This prototype, built using the shell of another VR headset, was only capable of rotational tracking. However, it had a 90-degree field of vision that was previously unseen in the consumer market at the time. In 2014, Facebook purchased Oculus VR for what at the time was stated as $2 billion but it was later revealed the more accurate figure was $3 billion.

By 2016, there were at least 230 companies developing VR-related products. Amazon, Apple, Facebook, Google, Microsoft, Sony and Samsung all had dedicated AR and VR groups.

With the COVID-19 restrictions in 2020, VR is experiencing an enormous rise. According to Grand View Research, the global VR market will grow to $62.1 billion by 2027.

VR is used across many applications such as cinemas, sporting events, and fine arts. It is so powerful that it can also replicates theme parks rollercoasters. In 2019 during a conference, I was invited to speak at (CEBIT 2019), I had the opportunity to test a VR rollercoaster. The visual, sensory, haptic experience was mind blowing.

VR is also used for therapeutic purposes. It is being used in the rehabilitation processes for elderly individuals that have Alzheimer's disease. This gives elderly patients the opportunity to simulate real experiences that they would not otherwise be able to experience due

---

35 https://en.wikipedia.org/wiki/Palmer_Luckey

to their current state. There are 17 recent studies with randomised controlled trials that show VR applications are effective in treating cognitive deficits with neurological diagnoses.

VR is also seen as an emerging technology in the commercial and diplomatic world. Brands are advertising and opening shops in virtual worlds like Second Life. Countries are opening embassies in Second Life. The Maldives were the first country to open an embassy in Second Life, followed by Sweden and Estonia in 2007, then Serbia and Colombia in 2008.

This trend signals the augmentation of our reality into digital realms. Imagine our digital twins will not only be capable of trading, completing tasks and solving problems your physical reality but also in digital worlds.

HOLOGRAPHIC TECHNOLOGY

Holographic displays uses light diffraction to create a virtual 3D image. The use of this technology was popularised in the collective psyche by *Star Wars*. In the George Lucas' directed movie, the technology is used by the *Resistance* to communicate and interact visually, in real time, across long distances.

Holographic displays are distinguished from other forms of 3D displays in that they do not require the aid of special glasses or external equipment for a viewer to see the image.

Virtual Reality development timeline:

The technology emerged in 1947 when Hungarian scientist Dennis Gabor[36] (Nobel prize winner in 1971) first came up with the concept

[36] https://en.wikipedia.org/wiki/Dennis_Gabor

of a hologram while trying to improve the resolution of electron microscopes. The name for holography came "holos" being the Greek word for "whole", and "gramma" which is the term for message.

In 1960, Russian scientists Nikolay Basov and Alexander Prokhorov[37], and American scientist Charles H. Townes[38] developed the first laser. This was a major milestone for holography because laser technology is the basis of some modern-day holographic displays.

In 1968, Stephen Benton[39] invented white-light transmission holography. This type of holography was unique because it could reproduce the entire colour spectrum by separating the seven colors that create white light.

In 1972, Lloyd Cross produced the first traditional hologram by using white-light transmission holography to recreate a moving 3D image.

In 1989, Spatial Imaging Group at the Lab MIT pioneered electroholography. It uses magnetic waves and acoustic-optical sensors to portray moving pictures onto a display.

In 2005, the University of Texas developed the laser plasma display, which is considered the first real 3D holographic display.

In 2010, hollow pyramid shaped prisms were released into the consumer market. If you placed them over a flat screen (or smartphone), it can emulate a 3D image using 2D light refraction.

In 2012, the first holographic display is implemented in a car's interactive navigation display system. The technology was showcased in the exclusive luxury car, the Lykan HyperSport.

---

37 https://en.wikipedia.org/wiki/Alexander_Prokhorov

38 https://en.wikipedia.org/wiki/Charles_H._Townes

39 https://en.wikipedia.org/wiki/Stephen_Benton

In 2013, MIT researcher Michael Bove predicts that holographic displays will enter the mass market within the next 10 years, adding that we already have all the technology necessary for holographic displays.

What used to be science fiction 10 years ago is now used at conferences and in entertainment shows.

In 2019, I attended a conference in Sydney, Australia where the hologram of a speaker located in London, UK was used to share insights, learning and experiences with the amazed attendees.

Dead musicians are now taking the stage with their holograms (digital twins). Buddy Holly, Roy Orbison, Tupac, Michael Jackson and Whitney Houston can now be experienced live in concert.

This technology not only brings the dead back and enables their IP owners to continue exploit the asset.

In that sense, it creates a fascinating opportunity to use your lifespan to train your digital twins to be the best they possibly can be in specific fields and once you pass on, the IP is transferred to your heir. If this happens in the future, it will be undeniable that digital twins will become one of our most valuable assets.

Holographic technology is not in its infancy and already presents possibilities for our digital twins to augment our physical presence anywhere in the world and in real time.

It can also be argued that our digital twins will become a pathway to immortality. Your intelligence, knowledge, experiences can be left to your kids and grandkids. It will become a means to be digitally present in their lives when your own ends.

Let us imagine a world where we could replicate the thought process of key persons of influence who guided humanity. Their digital twins could be trained to think and perform like them. We could bring these incredible minds back in to digital existence and have them continue shape our world.

Technology could bring back Mahatma Gandhi, Nelson Mandela, Winston Churchill, Victor Hugo, Voltaire, Leonardo Da Vinci, Pythagoras, Socrates. Their infinite wisdom could lead mankind in times of doubt. They could be advisors to the United Nations. They could be Digital diplomats working tirelessly to avoid conflicts. They could teach their wisdom in schools, universities. They could challenge populists and warn people about the danger they represent.

Imagine a world where the Masters of Renaissance continue producing art that shapes minds. Or a world where George Washington and Benjamin Franklin attend political rallies and continue to fight for democracy. Imagine a world where Keynes, Hayek and Schumpeter are lecturers in economics and students can interact with them. I would be willing to pay a lot to attend a class like that. Or imagine concert arenas in trance witnessing Freddy Mercury mastering his craft.

This leads us to the next enabling technology — digital humans.

DIGITAL HUMANS

Digital humans is an emerging concept that leverages certain technologies and aims to create a digital representation of a human being. It can be a real person or completely engineered.

In its current form, a user interacts with a digital human through a display. In the near future, interaction will be possible through AR, VR or holograms. Some of the technology used to create digital humans includes AI, NLP and DL. A leading company in this field is Soul Machines[40]. The Auckland-based technology company was founded by Greg Cross, a successful tech entrepreneur and Dr Mark Sagar, a double Academy Award winner and Director of the Laboratory for Animate Technology at the Auckland Bioengineering Institute. Dr. Sagar won the Academy Awards for his work on the animated movies *Avatar* and *Spiderman*.

Soul Machine developed the concept of digital brain in order to centre their research around progressing human-machine collaboration. By combining models of physiology, cognition and emotion, Soul Machine is creating new forms of biologically inspired AI.

The results are mind blowing. Soul Machine has the technology to digitise a real human (think about a world class movie effect) and enhance that human with AI that has a cognitive system. The machine is capable of recognising your mood by the tone of your voice and by processing your questions and comments. Not only will the digital human answer your questions in an intelligible manner but it will also read the micro expressions on your face and adjusts its own facial expressions to mirror your feelings to create a sense of ease. It does this by relaying information about your mood, intent, state of mind. Hiding these micro expressions is almost impossible to control.

Soul Machine's intent is to be the world leader in human-machine interaction.

There are multiple commercial applications for digital humans. Soul Machine helps large organisations in commoditised industries (banking, insurance, energy, telcos) interact with customers. Rather than walk into a store or calling a contact centre, this online technology

40 https://www.soulmachines.com/

helps customers navigate a product range or answer questions related to their product/service experience. In this scenario the digital human becomes the brand's ambassador. This signals a profound shift in both customer interaction but also in terms of trust proxy. If large brands like RBS, Mercedes-Benz, the New Zealand Police Force and many other adopting the technology are entrusting a digital human to interact with their most precious asset — their customers —it clearly indicates use of the technology is on the rise.

For the moment this technology is just used to showcase innovation and solve insignificant volumes of customers' requests, but the possibilities are limitless.

It is foreseeable that it will quickly scale and become a sub distribution and a customer relationship management channel.

The technologies we have talked about (and only scratched the surface of) are not only reshaping how organisations are interacting with their customers but they are also reshaping the workplace.

I used to lead the innovation team of a large financial services incumbent with hundreds of products in its portfolio. These spanned from a whole suite of banking products, property, car and business insurance. Most customer interactions were still managed through contact centres. No human in the world has the cognitive abilities to know hundreds of products by heart. This includes their regulations, intent, forms so it is close to impossible to give professional advice to a customer. You have to consider specific needs, risk appetite, investing profile, regulations across a range of hundreds of products and so on. But digital humans can!

Not only can a digital human learn all the parameters of products suites, they can also read human emotion and reply accordingly. Digital humans are also scalable. Their only limitation is the network bandwidth. Digital humans will likely start increasing their volumes of interactions and then will be used as a triage mechanism and first point of contact for businesses. By doing so, they should be able to handle the vast majority of low value or less complex customer interactions while real contact centre professionals handle the more complex queries. More complex in terms of technicality but also emotionally (e.g. complex claims).

Digital humans will be key enablers for digital twins. This technology will enable other people or organisations and/or their twins to interact with our digital beings. It will enable us to not only replicate our physical attributes (shape, face, an eye, hair, skin colour) but to also digitise our emotions. This will create genuine and trustworthy digital interactions.

THE MEMORY OF OUR DIGITAL TWINS

DISTRIBUTED LEDGER TECHNOLOGY OR BLOCKCHAIN TECHNOLOGY

Blockchain technology is also known as distributed ledger technology (DLT). This became a hot technology in the late 2010s early 2020s. The price of DLT backed assets like bitcoin are reaching all-time highs and tech start-ups are using the catch phrase "...and we use blockchain" in their pitch, to raise tremendous amounts of money.

At its core, DLT is an encryption mechanism. Information created or accessed is split into parts (known as blocks), encrypted, timestamped and attached to the next block. Theoretically this technology makes the registered information unalterable and, therefore, creates a safe, auditable record of activity.

A blockchain can be public such as major cryptocurrencies like Bitcoin or it can be private.

The blockchain was invented by a person (or group of people) using the name Satoshi Nakamoto[41] in 2008 to serve as the public transaction ledger of the cryptocurrency bitcoin. This made bitcoin the first digital currency to solve the double-spending problem without needing a trusted authority or central server. The bitcoin design inspired the creation of other applications and blockchains the public can read and are widely used by cryptocurrencies.

Cryptographer David Chaum first proposed a blockchain-like protocol in his 1982 dissertation *Computer Systems Established, Maintained, and Trusted by Mutually Suspicious Groups*.

In 1992, Haber, Stornetta, and Dave Bayer incorporated the Merkle tree into the design. This improved its efficiency by allowing several document certificates to be collected into one block.

41 https://en.wikipedia.org/wiki/Satoshi_Nakamoto

Nakamoto improved the design using a hashcash-like method to timestamp blocks to eliminate the need for them to be signed by a trusted party. This introduced a difficulty parameter to stabilize rate with which blocks are added to the chain, therefore, making it more secure.

In August 2014, the bitcoin blockchain file size that contains the records of all transactions on the network reached 20 GB.

In January 2015, the size had grown to almost 30 GB, and from January 2016 to January 2017, the bitcoin blockchain grew from 50 GB to 100 GB in size. The ledger size had exceeded 200 GB by early 2020.

One of the powers of blockchain is that it creates unalterable evidence of provenance. By proving provenance (identity, intellectual property, transaction, etc) you demonstrate trustworthiness. As we saw earlier, trust is the currency for interaction.

Blockchain will enable you and your digital twins to prove who you say you are. It will also guarantee third parties that the intellectual property used in interaction is fully owned and not in breach of someone else.

We have just explored some existing technologies that will become key enablers for individuals and organisations to leverage the power of digital twins. To some extent, organisations are already adopting the technology to replicate attributes of key assets such as machines, intellectual property and processes.

We explored, through examples and scenarios what is possible given the existing and accessible technology as of the early 2020s. If you think that you just read a utopic science fiction story, bear in mind that it is not. And emerging technologies will accelerate the capabilities of digital twins even more. That is the intent of the next section — quantum computing.

THE IMPOSSIBLE BECOMING POSSIBLE – QUANTUM COMPUTING

Quantum computing is the use of quantum phenomena, such as superposition and entanglement, to perform computations. Quantum computers are believed to solve certain computational problems, such as integer factorisation (which underlies RSA encryption), substantially faster than classical computers.

Quantum computing began in the early 1980s, when physicist Paul Benioff[42] proposed a quantum mechanical model of the Turing machine.

In 1994, Peter Shor (Professor of Mathematics at MIT) developed a quantum algorithm for factoring integers with the potential to decrypt RSA-encrypted communications.

In recent years, investment in quantum computing research has increased in both the public and private sectors. On 23 October 2019, Google AI, in partnership with NASA, claimed to have performed a quantum computation that is infeasible on any classical computer.

Quantum computers are emerging at scale and will dramatically increase computation power and exponentially enhance the speed your digital twin will learn, interact and grow at. It is even likely your digital twin will outsmart you (quite quickly). It may reach conclusions to problems your mind is not geared to even conceive.

As you may imagine, the first industries to heavily invest in quantum computing are the tech giants (Google, IBM). Quantum computing will accelerate the race toward AI dominance.

---

42 https://en.wikipedia.org/wiki/Paul_Benioff

The car manufacturing industry also is looking at this technology to create AI enabled self-driving vehicles.

Banking giants such as Goldman Sachs were among the first to invest in quantum computing technology. And this computing power will exponentially improve financial modelling and provide a competitive edge.

Quantum computing will enable highly complex encryption models for the military and cyber security.

The pharmaceutical industry is willing to invest in quantum computing for molecule development.

Quantum computing is not expected to be at scale before the mid-2030s. So, it leaves roughly a decade for incumbents across the finance, car manufacturing, pharmaceutical and other industries to start thinking about how to prepare their businesses to leverage this emerging technology.

APPLYING DIGITAL TWINS TO FINANCIAL SERVICES

HOW DOES THE CONCEPT OF DIGITAL TWINS APPLY TO FINANCIAL SERVICES?

Now that we have introduced the concept of digital twins and the technologies (existing and emerging) that underpins its development, let's explore how it could apply to financial services.

The focus of this book is on financial services but given it is a commoditised industry, the reader can apply the same principles to industries like energy, telcos and contextualise.

At its core principle, financial services provide commodities to its customers. The main commodity is money. For a profit, a bank will give its customer access to liquidity to procure a valuable asset at a static point in time. This asset can be physical such as a property, vehicle, equipment, machinery or digital (software).

In order to deliver that money to a customer, the bank relies on a series of strategic assets including:

- Access to capital from central banks through a charter and licence.
- Brand, reputation and perceived trustworthiness to attract customer.
- Core IT infrastructure and support functions.
- Distribution channels (digital, physical, intermediated).
- The people who interact with customers.

If we consider the economy has a human body, banks are the heart, pumping the blood (money) and distributing it to the different organs that make up the economy.

On the back of the banking system, another key player in the industry are insurers. Whether they are categorised as general, commercial, specialised, health or life, insurers provide protection for customers' valuable assets.

Using the same analogy, if banks are the blood that irrigate and nourish the organs, insurers are the bio defence against threats to the body and its organs. Insurance acts as a shock absorber for the economy, through the mechanism of risk pooling and, against a premium, an insurer will remediate damage that occurs to an insured asset.

RISK UNDERSTANDING: CORE CAPABILITY OF THE FINANCIAL SERVICES INDUSTRY

At the core of the financial services Industry is the notion of "risk". The core capabilities the industry has built are geared to understand the risk profile of an asset, to understand its value, how the asset will be used by the customer base and how to finance and protect it.

Today the industry faces a key challenge to transform itself as fast as society is transforming.

The banking industry as we know it dates back to around 1130AC. The Poor Fellow-Soldiers of Christ and of the Temple of Solomon also known as the Knight Templars invented the system. Created by Hughes de Payens[43] and eight other knights, their mission was to protect the Christian pilgrims on their journey from Europe to Jerusalem. On their way they were subjected to violent attacks exposing them to robbery but also threats on their lives. The Knight Templars developed a network of fortified places along the pilgrimage route to Jerusalem. This was the first type of a bank branch network to protect and make

43 https://en.wikipedia.org/wiki/Hugues_de_Payens

pilgrim's money available along their journey. It was an early form of risk management for the pilgrims money and their assets.

The Knight Templars became so efficient at it that their wealth started threatening the Pope and the King Philip IV of France[44], who, deeply in debt, saw an opportunity to claim their wealth to repay his debt.

A PARADIGM SHIFT: THE RISE OF DIGITAL ASSETS

One can argue the financial services industry still works on the same principles that it did eight centuries ago. Banks give access to money all over the world and insurers protect the assets you buy with the money a bank provides.

However, we are experiencing a fundamental shift of paradigm, accelerated by technology and societal changes. The shift is focused on asset value and its perception in the eyes of customers.

The financial services industry is geared to provide customers the means to procure physical assets. For example, a bancassurer has a value chain geared to enable the acquisition of physical assets. That very organisation wants its customer to buy a car or a property so it will give the customer access to a mortgage or a loan. As these physical assets cannot be easily replaced, there is a need for protection mechanism through insurance. A house is one of the most expensive assets you can own and serious uninsured damage can cause you financial distress.

44 https://en.wikipedia.org/wiki/Philip_IV_of_France

Some insurers have even vertically integrated the value chain. They own large construction companies that can complete repairs. They own networks of smash repairers to repair cars.

If that integrated model makes sense from an economical perspective by the enablement of economies of scales, claims processing and cost efficiency, it also presents a full risk exposure to a paradigm shift. The integrated model is highly efficient where the core business assumption is that "people will own and value physical assets".

Now if we shift the assumption to "people will own and value digital assets" the entire value chain as we know it collapses and will leave banks and insurers struggling to transform.

I am going to reiterate a previous example I use to illustrate that point. As an individual who spent his first three decades in France, I bought a property in Paris in 2009. Should I sell the property today, I would make a nice capital gain of about 25 – 30%. If in the same year I had invested $100 in Bitcoin, at time of writing this book, I would be a billionaire. Now let's assume that I simultaneously hold these two asset classes (one physical and the other digital), which one do you think I want to protect as a top priority? My apartment or my crypto wallet?

As far as I wish I had billions worth of cryptocurrencies, to my knowledge and at time of writing, no financial institution in the world can provide me with insurance or finance for my digital asset. However, there are countless banks and insurers who can provide me with undifferentiated services for my mortgage, home insurance and landlord insurance.

This paradigm shift from physical to digital is not science fiction. It is backed by numbers and we'll explore the consequences.

Volumes of new individual vehicles sold[45] fell across the world, accelerated by the Covid-19 pandemic. European car sales fell 24% in the year to January 2021. US data shows car sales at minus 30%[46] in June 2020, as emerging generations embrace mobility services.

Access to home ownership is becoming more and more difficult as prices peak, forcing entire generations to rent.

The Quantitative Easing policy from Central Banks (FED, ECB, BoE) enable access to almost free money. This monetary policy had a side effect of inflating the value of assets like realestate. Baby boomers can leverage their property portfolio to borrow more and increase their position. This is good for them but terrible in the short term for their kids and grandkids.

These trends are threatening the financial services industry which has built its appreciation of "good risk" on the template of Baby Boomers being hungry for physical asset ownership.

EMERGING NEEDS AND GENERATIONAL CONFLICT

Research suggests that emerging generations are more prone to asset usage than ownership. A Stanford Centre for Longevity study[47], highlighted that home ownership was at its lowest in 50 years. This trend was accelerated by the digital economy and some service subscription models.

---

45 https://www.bloomberg.com/news/articles/2021-01-19/europe-car-sales-drop-most-on-record-in-year-bedeviled-by-virus

46 https://www.reuters.com/article/us-usa-autos-outlook-idUSKBN23W2OJ

47 https://longevity.stanford.edu/home-ownership/

Unlike their parents who grew up in a standardised consumerism driven society, Millennials, Gen Z generations are digital natives. They were born or raised with access to technology and their habits have been shaped by the use of computers, smartphones, hyper connections and unlimited access to digitised information, knowledge and IP.

The financial services industry needs to adapt to that emerging generation and the ones that follow. Not only are they currently either underserved or poorly served, but they will be the recipients of the largest wealth transfer the world has ever experienced.

In the US alone, it is expected that baby boomers will transfer $30 trillion[48] in wealth to younger demographics over the next 10 to 20 years. This receiving generation will likely:

- Be attracted to digital assets. Banks and insurers are on a burning platform and will have to adapt.
- Also be the first to augment themselves with digital twins.
- See value in financial services proposing services for physical assets based around digital twins.

They may inherit a property from their parents but it needs maintaining. The maintenance history could be digitised and the information used to better price the risk. It will enable an accurate valuation of the property. And from that valuation, an insurance premium can be derived.

The financial services company that creates a digital twin of an asset will have a strategic advantage over its competitors. It will have the information to create an ecosystem of services designed to serve the exact needs revolving around the asset. For example, when knowing the

48 https://www.forbes.com/sites/markhall/2019/11/11/the-greatest-wealth-transfer-in-history-whats-happening-and-what-are-the-implications/?sh=571060424090

plumbing is 40 years old, insurers can tailor risk models. In this case there is the likelihood of water related damages which can lead to preventive maintenance services. Knowing a property is not energy efficient could lead to building models and propositions around energy efficiency and upgrading to renewable energy or becoming carbon neutral.

WHEN LOVE, SECURITY, CONTROL, AND POWER DRIVE YOUR RELATIONSHIP WITH MONEY

If a financial services provider earns the trust of its customer, it will gain the right to play with Open Banking and create digital twins of its customers by analysing of their transaction data.

Open Banking is a regulation (in the UK, EU and Australia) that forces banks to give access to customer's transaction data to any third party that can prove it has the customer's authorisation.

With data being the "new fuel", the regulator's intention is to foster competition through the development of new services that leverage transaction data.

It is possible to determine a psychological profile of someone by analysing their consumption. "Tell me how you consume, and I'll tell you who you are."

In a way our bank transactions reflect how we perceive and consume the material world that surrounds us. Our transactions are a form of digital twin that banks and insurers can leverage to create dedicated offerings.

In a 2018 survey by SelfHackathon[49], Dr Patrycja Slawuta and her partner Edward Loh analysed the psychological relationships of 5,000 participants with Money. From the study 16 profiles emerged, which were condensed into four major traits: Security, Love, Freedom and Power.

- **Security money profile.** Conceptually people with a security money profile are more likely to see money as a means to secure a way of life and would have a conservative profile.
- **Love money profile.** People with a love money profile are more likely to associate money with being loved.
- **Freedom money profile.** People with a freedom money profile are more likely to associate money as an enabler to break free of the bonds of employment.
- **Power money profile.** People with a power money profile are more likely to associate money with enabling dominance over others.

Based on the money profile and transaction analytics, digital twin patterns emerge. And by offering tailored financial solutions to customers, there is a better chance of closing the deal rather than using traditional marketing approaches.

By anticipating customers' needs through spend and psychological profiling, incumbents will be in a position to create solutions for their customers. These solutions will go beyond the traditional core suite of products and will enable the incumbent to grow through adjacencies. They will earn the right to play a facilitation role in their customers' lives. They will also be opportunities to (re)gain trust by extracting value for their customers not from them.

This leads us to the fundamental concept of trust we discussed earlier.

---

49 https://www.selfhackathon.com/

EMERGING FORMS OF TRUST: DIGITAL REPUTATION – THE 'TRUSTBOND' USE CASE

DIGITAL TRUST, AN EMERGING ASSET CLASS

As we have seen in earlier chapters, profound shifts are occurring to fuel the rise of digital twins. Societal shifts like worldwide adoption of all things digital and new ways of working (the gig economy and P2P) are reshaping relationships between employers and employees. New ways of consuming through accessing assets rather than owning them have emerged.

Technology is accelerating these shifts but when we explore the fundamental concept underpinning all interactions, it is "trust" which is at the centre of the game.

We saw how platform business models like Amazon, eBay, Facebook, Airbnb, Uber and Freelancer.com enable a per-to-peer rating system. We saw how tech start-ups like Traity and Reputationaire enable users to leverage their digital footprint or use their digital twins to prove trustworthiness to gain access to services.

Let's dive into a real example of how digital reputation carries tremendous value and why it can be seen as an emerging asset class.

In 2017, Australian bancassureur Suncorp had its strategic innovation team explore the concept of digital assets. The team took a deliberate approach to exploring digital reputation as an emerging asset class in the digital economy.

After extensive research the team decided to partner with Spanish start-up Traity. Over the years, Traity built and trained an algorithm that had a community of four million participants. They had empowered people to achieve a digital reputation score. Users could authorise Traity to access their digital public profiles such as LinkedIn, Airbnb, Amazon, Uber and eBay. Traity's algorithm then aggregates the different scores received by the user on these platforms and presents it in a form of consolidated "star rating".

As an example, should I granted Traity access to my digital profiles, it would return a result of 4 or 5 stars. So what?

The fundamental value of this capability is to prove trustworthiness. If someone is trusted by people, they do business with online, and they leave a positive review, it should signal a high level of trustworthiness. That trustworthiness is core to the notion of "risk" which is the cornerstone of the financial services industry.

With that capability, Suncorp designed an experiment which would span across the McKinsey 3 Horizon[50] framework.

- Horizon 1 (H1) captures the opportunity to enhance the business of today.
- Horizon 2 (H2) captures the opportunity to grow through adjacencies.
- Horizon 3 (H3) captures the opportunity to renew a business model.

Suncorp's intent was to leverage its strategic assets and the newly accessed digital reputation capability to explore the three horizons.

The team deliberately anchored its design and experiment in H3 and created a pathway back to H1.

---

50 https://www.mckinsey.com/business-functions/strategy-and-corporate-finance/our-insights/enduring-ideas-the-three-horizons-of-growth

By focusing on H3 you construct propositions anchored in emerging needs and, therefore, free yourself from the constraint of the system you operate within (refer to Chapter 1: why long term thinking and innovation matters).

The H3 ideation was anchored in the need to understand digital assets as an emerging asset class. As we discussed earlier, if customers start valuing digital assets and banks and insurers are only equipped to finance and secure physical assets, they will not be equipped to have a resilient business. The research indicated that digital reputation is an emerging asset class. Like any other asset class, at scale it will need to financing and protecting. The team wanted to prepare the organisation for such a world.

The H2 adjacent opportunity dictated to position an experiment in a new market, untouched by financial services. The research drew the team to the Australian rental cash bond market.

As we can see, an entire generation of Australians are locked out of the property ownership market. They have no choice but to rent (this is not unique to the Australian real estate market). Cities like Sydney and Melbourne scores high in the world rankings of most expensive real estate market. The competition is fierce to rent decent properties. Rental agencies push forward tenancy applications that fit an established framework that is built to demonstrate trustworthiness and financial stability.

This framework was built when direct employment was the norm and a single source of income proved financial stability. This framework doesn't recognise the rise of the gig economy, where an individual can earn the equivalent of a more conventional income but through hundreds of interactions. An Uber driver, a Airbnb host and trading items through an Amazon online shop will generate revenue from multiple sources. According to the current risk rating system, this sort of person is considered a "bad risk" and is penalised when trying to rent a property (or access financial services).

It's the same for a migrant who may be qualified in their field and trustworthy but doesn't have a credit or rental history in their new country. Reputationaire was founded on this basis. Its cofounder migrated from India to Australia with a master's degree in computer science and a job secured in Melbourne. Not fitting the standards, after many refusals, she decided to print her Airbnb reviews which were accepted by rental agencies and landlord as proof of trustworthiness.

On top of being a competitive market, renters have to pay one month's rent in the form of a rental cash bond. This bond is held by a third party (generally speaking a public entity) and is only released after the tenancy ends, if the property is left in good condition. This amount of money is significant as it can be a few thousand dollars. The rental cash bond market in Australia worth $4bn and is untouched by the financial services industry.

However, this poorly met need presents a significant opportunity to solve renters' problems. Rental cash bonds are an inefficient form of insurance.

To address that pain point, Suncorp and Traity developed a product called "Trustbond[51]". The product was designed to enable pre-establishing trust between a renter and landlord. The applicant would use the Traity star rating to prove their digital reputation and Suncorp issued an insurance product with the same value as the traditional cash bond. The renter only paid a small upfront fee and should damage occur during the tenancy, the landlord could claim the insured amount to cover any repairs.

[51] https://www.suncorpgroup.com.au/news/news/innovation-rewarded-for-building-trust

This product enables renters to free up their money at a point in time when they need it. Moving property is an expensive exercise. It also gave Suncorp the opportunity to create an ecosystem of services relevant to the journey of renting a new property.

Not only does it address a significant pain point, it also helps people keep their financial wellbeing on track. Australians only have little savings and a significant portion of the working population lives from one pay cheque to another. An outdated system like rental cash bonds is a double penalty for these people and those who don't fit the current framework.

The product was digital end-to-end and promoted through Suncorp's landlord insurance channel "Terri Sheer[52]". Suncorp received the Canstar 2018 Innovation Excellence Award[53] for its Trustbond product.

By developing Trustbond, Suncorp's vision was to position itself in a $4 billion untouched market as well as being an enabler of trust by empowering people to use their digital footprint for their benefit.

The H1 play was to use Suncorp's core capability and assets, as an insurer, to develop a fit for purpose product.

What we learned from this product development and launch is that digital reputation is an emerging asset class. Not just a hypothesis. In fact, digital reputation will likely be a core enabler behind our digital twins to generate trust to transact with other digital twins or humans.

A positive digital reputation will accelerate new revenue opportunities. The more our digital twins are trusted, the more engaged they will

---

52 https://www.terrischeer.com.au/

53 https://www.suncorpgroup.com.au/news/news/innovation-rewarded-for-building-trust

become. This core capability will be our brand and need protecting from threats such as cyberattacks, identity theft, allegations and false claims. Banks and insurers will have a role to play in protecting our digital twins and digital reputations.

We have just seen an example of how emerging forms of digital twins are already being leveraged in the financial services industry.

In the next chapter we explore the potential applications for customers, incumbents as both providers and employers. We'll also discuss the societal challenges that digital twins will create.

IMPLICATIONS FOR FINANCIAL SERVICES

As we discussed earlier, at scale, digital twins will be strategic assets for their owners and users.

Let's now explore the applications for customers, financial services incumbents, employers and the societal changes that will need to occur.

OPPORTUNITIES FOR THE CUSTOMER AND ASSET OWNERS

EXPANDING THE NOTION OF TIME

When you think about the possibilities explored earlier in this book, you have probably wondered how the ability to replicate yourself through a digital twin will change your life.

The choice will be yours, but you are likely to be able to delegate low value/high volume mundane tasks to your digital twin when in its infancy. Tasks like paying your bills, managing your calendar, writing emails on your behalf, doing your tax, managing your expenditures and keeping you accountable to your health and fitness goals.

It is very likely that you will use your digital twin as an advanced personal assistant. This will free up time to focus on more meaningful tasks and your lifestyle. Perhaps you will use this time to reinforce your family, social and professional bonds. Maybe you will spend more time augmenting your learning, spirituality or working to strengthen the core pillars of your identity.

A NEW FORM OF INTELLECTUAL PROPERTY TRANSFER

As your digital twin demonstrates its ability to deliver efficiently as your ambassador, you will grow to trust it. Once trust grows, you will spend more time training it. As trust grows, so will your self-trust. Training your digital twin may also make you realise there are aspects of your life you want to enhance or change. As an example you could realise that you lack technology related skills and that you want to learn coding. You could decide that your personal edge is related to your interpersonal skills, as a result you could invest in training yourself to then make your digital twin even more personable.

The longer you work with your digital twin, you will soon start transferring more valuable tasks, like handling interaction with your employer(s), accountant and lawyer, to your digital twin.

The more you reinforce your IP, the more you will accelerate the enablement of your digital twin to represent you and generate more revenue for you.

Bear in mind that your digital twin never sleeps or rests and is not constrained by physical and geographical boundaries.

AN EMERGING ASSET CLASS AND THE MAIN SOURCE OF REVENUE

Your IP is an emerging asset class. The more your digital twin performs based on your IP, the more confidence you will have in your digital twin handling hundreds of simultaneous engagements, across the world, on countless different platforms, in different languages.

Your digital twin, at some point, will become your main source of revenue and become a pillar to your life's stability. You will need to find mechanism of protection for that asset. Perhaps that service provider will be a bank or an insurer, perhaps it will be a Tech giant. At time of writing this book I can't foresee banks and insurers being ready to position themselves in the emerging space of Digital Assets protection and financing. This space represents a significant growth corridor, and should financial services incumbents decide to explore that space, they will position themselves in the infinite game.

OPPORTUNITIES FOR THE INCUMBENT

Digital twins will force incumbents to rethink multiple layers of their organisation and the capabilities they leverage to interact with their target audience.

CUSTOMER INTERACTION

Traditional communication/interaction channels are less and less relevant with hyperconnected customers. For example:

- Advertisements on broadcast television are cannibalised by the hyper targeted capabilities of the likes of Facebook, Instagram and Google.
- The consumption of streaming services like Netflix, Amazon Prime and Spotify make no room for commercials. In fact, being commercial free is probably one of the characteristics that appeals to customers.
- Digital channels (websites, Progressive Web Apps) shrink foot traffic in prime real estate branches.
- Contact Centres will slowly but surely become a support function to digital channels.

In such a world, where digital twins are emerging, the cost structure associated with maintaining expensive assets such as a physical network is challenged.

THE RISE OF CONVERSATIONAL COMMERCE

Conversational commerce consists of building a presence through widely adopted messaging apps such as WhatsApp and Messenger. The technology uses chatbots with the ability to process customer questions. And if the question is too technical, it triages the request to a real human who will take the lead. This technology blends human and machine capability to service customers. Large incumbents in commoditised industries have already adopted that technology (Telcos, banking, energy and retail).

The technology makes the brand visible in a specific channel, aligned with consumer habit. It can also process large volumes of low value requests.

This technology is so powerful that an Italian neobank, Buddy Bank[54] (powered by giant UniCredit) uses conversational commerce as a primary customer interaction and communication channel.

A leading technology company in the field of conversational commerce is LivePerson[55]. (Their solutions are used by major incumbents operating in commoditized industries).

If we take the example of Buddy Bank, the conversational AI capability can be assimilated as a form of digital twin. In that scenario, the bank's digital twin will interact with the human customer. It is not a stretch to forecast that the customer's digital twin will soon interact directly with the bank's digital twin or a human representative. In scenarios

---

54 https://www.buddybank.com/

55 https://www.liveperson.com/

where the transaction becomes more complex or valuable, the digital twins will facilitate human-to-human interaction.

RETHINKING MARKETING AND COMMUNICATION STRATEGIES

It is also possible that entire marketing and communication strategies will be built to appeal to your digital twin. That would open an entire new domain for communication agencies and marketing departments. If your digital twin is an extension of you but your brain or attention is not available to view marketing campaigns, then the communication battleground will shift to your digital twin. It will make decisions based on what you taught it. And this removes the emotion out of making a purchase and helps keeps your finances under control. This would be particularly useful for securities trading. As your digital twin will be hyper vigilant and trained to place orders aligned to your risk appetite, it is likely to stop passionate or compulsive high-risk investments. A perfect example is cryptocurrencies which is (at time of writing) reaching an all-time high with some "coins" having unclear value propositions or underlying assets.

UNDERSTANDING THE RISK AND TAILORING THE RIGHT OFFERING

Your digital twin will hold your risk profile. Your digital twin will replicate who you are, your values, your thought process, your appetite for risk. It will reflect the core risk you represent by the way you make decision. Your digital twin will also enhance your risk profile. Your digital twin will be capable of removing emotions from transactions and will make better decision for you.

By absorbing our identity through the number of tasks they operate on our behalf and the digital reputation they receive as a form of customer feedback, digital twins will offer invaluable information about their "human sibling". This will enable the creation of dedicated solutions and perfectly match both our needs and risk appetite.

Organisations that fail to match your exact need will simply be blocked and triaged by your digital twins.

RETHINKING ENTERPRISE ARCHITECTURE

Still many incumbents across commoditised industries are product centric. They have built digital and physical fortresses solely dedicated to building products to push to market.

By organising their enterprise with this approach, customers are by default considered an attribute of the company's products and services. This creates a major risk exposure in a world where consumers have been used to hyper personalisation from tech giants like Google, Amazon, Facebook and Apple.

This product centricity mindset creates a legacy architecture that does not meet customers' who want solutions to matching their lifestyles and aspirations.

Digital transformation programs are in place in all major product centric companies to help them make the paradigm shift to treating their products and solutions as attributes of their customers.

Large players are being disrupted by digital first emerging players. They are being forced to play catch up to slow down customer churn.

Players like Zhong An[56], a Chinese insurer offshoot of insurer Ping An and Alibaba, has sold 600 million policies in three years since inception. It is arguable that China is a very specific market because

---

56 https://www.zhongan.com/corporate/?lang=en

of its scale but digital players like Zhong An have unlimited scalability and are serious threats to traditional incumbents. A threat serious enough for Japanese insurer Sompo, to buy Zhong An technology and replicate it to keep domestic market dominance.

This example is not an anomaly. It is a rule. All incumbents unwilling or too slow to transform will be doomed to irrelevance. If they can't handle their transformation, how will they prepare themselves to stay relevant in a world where consumers widely adopt digital twins as extension of themselves?

RETHINK THEIR SERVICE OFFERING: THE RISE OF CYBER INSURANCE

As we saw, financial services are geared to develop, distribute, process products and services designed to enable customers to access ownership of physical assets. It is likely to be less relevant in a world where digital twins and other forms of digital assets are valued by customers and are at scale. Instead services and products will need to prevent threats, protect assets from threats and remediate a damage to digital assets. This will be a critical for a competitive advantage.

The best example is cyber insurance. These are products designed to remediate a loss following a cyberattack or cyber hazard. This form of insurance is in its infancy as the risk is extremely complex to understand and so it is difficult to price. It is also very contentious to remediate.

According to a 2019 report from Munich Re (one of the world's largest reinsurers), the cyber risk (measured in terms of economic losses) was $US450 billion in 2016. It is projected to reach $US6,000 billion by 2021. This is a staggering 1,387% increase in five years. The greatest strategic risk for insurers is not offering fit for purpose solutions for their customers.

Huge companies with IT departments and cyber specialists with material cyber security budgets have recently been victims of cyberattacks. In 2011, RSA SecureID, a world leading company in the space of cybersecurity was infiltrated. In 2016 Yahoo lost 500 million records. In the same year the Bank of Bangladesh experienced a cyberattack and lost $81 million. In 2018, Twitter lost 330 million records, Facebook 90 million records and Marriott 500 million records.

These examples show that risk is growing quickly and that even large organisations are vulnerable to a constant evolving threat.

The most interesting recent use case that illustrates the strategic risk for the insurance industry is NetPotya[57] when it was a victim of a cyberattack (ransomware) in 2018. This attack targeted Ukraine at the time Russia was taking control of Crimea. Nuclear plants were hacked to destabilise the region. The sole fact that a nuclear plant was hacked, in the same country where the Chernobyl nuclear meltdown happened in 1986, is testimony to the potential impact of cyber threats.

NetPotya generated collateral damages (estimated at $10 billion). One major impact was interrupting business for confectionary giant Mondelez.

One of the Mondelez plants had to stop. Insured against "business interruption" by giant incumbent Zurich, Mondelez filed a $100 million claim under their business interruption policy. Mondelez was allegedly covered against cyber risk.

Zurich started investigating the claim and this was difficult due to the nature of the threat and insurance industry's difficulty estimating risk and pricing it accordingly. In parallel to the investigation, the USA

---

57 https://www.insurancebusinessmag.com/au/news/cyber/cyber-risk-knows-no-boundaries-161650.aspx

and Israel made separate public statements that NetPotya was an act of cyber warfare from Russia against Ukraine.

Zurich denied the Mondelez claim as losses due to act of war are a general exclusion for standard policies. The case has been brought to justice and is pending a verdict.

This illustrates the complexity of cyber insurance and the industry struggle to understand this fast-emerging form of digital risk.

Banks may also ask for insurance products to finance digital asset classes. In the same way they currently ask for Lenders Mortgage Insurance to protect themselves from customers defaulting on their mortgage. So it is likely they will want similar mechanism to protect their exposure to default on repaying loans for digital assets.

Insurers will only have a right to play to compete in the cyber security space by understanding the emerging forms of risks it present and developing tailored solutions. If they don't, technology giants will become legitimate players as cyber protection could be an adjacent play. To illustrate this, ask yourself who from Symantec, Norton or Axa, and Allianz would you trust to propose you cyber risk protection and remediation solutions.

We are starting to see similar plays in the automotive industry. Tesla has announced it will launch an insurance product. It will leverage the telematics onboarded in their vehicles to understand individual driver risk profiles and provide them with a tailored insurance product.

Tesla's adjacent play is legitimate and threatens the core business of general insurance.

THE END OF BRICKS AND MORTAR?

Covid-19 accelerated the adoption of digital as a main channel to use for interaction. Probably sentencing organisation with a large physical footprint to irrelevance. Banks, insurance companies and Telcos who built a physical presence over decades, now have to maintain an expansive and lesser strategic asset (it can also be seen as a strategic disadvantage).

However, they could use their prime locations to showcase technology advancements in their products and services. They can leverage the technology we explored previously to create accessible places for digital twins to interact. They could leverage holographic technology overlayed with an AI and digital human interface to answer any potential customer questions on the spot.

OPPORTUNITIES FOR EMPLOYERS

Most employers in commoditised industries approach human resources as the act of procuring time from workers with the relevant skills at a point in time to operate a specific job in that time. They buy those skills, on a time basis, in exchange for a salary. That transactional relation employer-employee has been the dominant model for more than a century.

From an accounting perspective, human capital is captured as a liability rather than an asset. Although it makes sense according to accounting standards, it poorly describes and gives credit to the power and abilities that people sell their employers.

With the accelerated adoption of technology and automation, workplaces are changing fast. Robotics process automation completes human tasks faster and more efficiently without the need to take a break.

Jobs that require a low level of emotional quotient to perform them are doomed to automation.

What we are experiencing is not new. Innovation has often rendered jobs irrelevant. Dactylographers were replaced by printers, telephone operators have long been replaced by machines, for example.

Innovation may threaten people's jobs, but it also provides a significant opportunity to upskill. The rise of digital twins is no different. It is an opportunity to enhance workers.

Let's take the example of a contact centre professional. That person has been trained to handle conversation with customers. They receive training so they can guide, inform and assist customers with most common requests on the most common products or solutions the company provides. In a high volume/low value call pattern, it is foreseeable that this type of roles will be disrupted and automated. It is already largely the case.

However, humans cannot yet be beaten by a machine when it comes to emotional intelligence. For the time being we, humans, remain the best at closing sales, solving conflicts, providing active listening and emotionally engaging.

MANAGING A HYBRID WORKFORCE

Hybrid workforce models are emerging where the machine augments the human and can provide career development pathway.

Let's take again the contact centre professional example. A large financial services company can offer hundreds of different products, is subjected to countless of regulations, different customer life stages, investment profiles, asset classes, and the list goes on. It is simply impossible for human cognition to professionally advise a customer on each of the potential scenarios driving their need.

The human brain, in a normal state of consciousness, can assimilate and simultaneously connect together between seven and nine variables to solve a problem. Human subconscious can operate using a far wider number of parameters simultaneously. After all, it is responsible for operating your body to keep you alive. The only problem is that accessing an altered state of consciousness is not a capability most of us can turn on or turn off.

Machines can handle immense numbers of variables simultaneously but are not good with emotions or communication. Due to their processing ability, AI is already enhancing human capital in managing complex suites of products and services to advise end customers.

As an example, software company IPSoft developed "Amelia[58]", a "digital colleague" does just that. Their AI is trained on complex and various product ranges to provide a safety net to a human working in a contact centre.

This leads us to an interesting concept of managing a hybrid workforce made up of humans and machines.

Currently companies are licencing software to access AI capability. They entrust the machine with the company's IP so the machine can replicate internal know-how and apply it in a specific context. This form of IP licencing is captured in well-established and known contractual mechanism binding a client-vendor relationship.

---

58 https://amelia.com/amelia/

On the other hand, although the concept is similar, the employer-employee relationship differs. The employee provides time and experience and intelligence to apply the employer's IP in a relevant context. But that contractual relationship is bound by specific employment laws that are different from one state to another. On top of the contractual relationship, an emotional relationship is created.

Employees are also not scalable/replicable (yet), where a software expansion is limited to the number of licences procured.

A PATHWAY TO UPSKILL THE WORKFORCE

A pathway for employers will be to upskill their workforce by giving them access to technology so they can create digital twins. Employees could replicate their knowledge and emotions and add it as an overlay to the company's IP. This would create a diverse digital human workforce, capable of answering any customer requests professionally.

Let's imagine this contact centre professional that can advise customers on between five and 10 major products or services. Having a digital colleague assisting with access to the entire company's knowledge could, over time, could create an internal career progression pathway for humans. As the digital colleague gets to know the human professional better, the machine could also advise on a tailored, optimised career pathway based on the person's emotional intelligence, preferences, performance, aspirations and psychological profile.

One of the key questions to solve will be compensating employees. If employees agree to digitally replicate their knowledge and emotion, it will be a form of IP licencing and they should be compensated for this. Imagine a mechanism where the most performant digital twin will deliver an incentive to its human sibling after each transaction. A digital human leader board could be created and used to understand the performance patterns used to upskill all humans and digital workers.

It is possible, in the near future, employers will have to train employees to manage a workforce made up of humans, digital humans and machines. This will require specific management skills as well as specific contractual arrangements between employer/employee; vendor/customer; and vendor/employee.

We have just explored how pathways to upskill workforce can be achieved. It is one example out of many options.

WHAT DO I DO WITH MY PEOPLE?

A more disturbing question is arising when an upskilling scenario isn't possible. The question is, what do we do with our workforce? Can we offer them different career pathways within the organisation? Do we have to make most of them redundant? More than ever the philosophical opposition between considering human capital as an asset or a liability will be challenged.

That question will keep leaders, executives and board directors up at night. The answer will depend on the context but also on how well prepared the organisation is to embrace the concept of digital humans and digital twins.

OPPORTUNITIES FOR SOCIETY

We have explored, and only scratched the surface of how deeply digital twins will shape the future, in terms of:

- Shifting the perceived and established value of physical assets toward digital assets.
- Changing how digital assets like plants, mines and planes are operated and maintained.
- We discussed the key enablers from a technological and psychological perspective.

By now you should have a pretty good understanding of the magnitude of the change that is coming.

But the rise of digital twins will force our society to make significant decision about the path we want to explore. Digital twins will likely create even more knowledge, wealth disparity as well as power concentration. This topic will need addressing through public forums. Politicians should start upskilling themselves to understand what is at stake.

Digital twins presents both the ability to empower humans with their knowledge, traits, experiences and expertise and to augment themselves through unlimited ability to download external IP. It also represents a concentration of risk. Only a few companies have the technology to connect all the enablers for digital twins to scale. Those companies are not philanthropic so it is difficult to imagine that they will empower people to free themselves from the traditional bonds of employment.

It is utopic to believe that employers will choose to upskill their workforce and maintain extended or outdated knowledge and delivery power when machines can do it better and faster at a fraction of the cost.

Societies faces the threat of major unemployment due to the incapacity of the education system to pivot to meet changing needs.

This presents a risk to democracy because when large amounts of people struggle to survive, uprisings are never far down the track. The French Revolution is a perfect example. Millions left starving to death because of an unbearable tax burden and a concentration of wealth in the hands of very few.

Today's situation is not dissimilar — 3% of the world's population controls 95% of the wealth.

People without work for long periods of time will either lead to mass reproduction, mass killings or a mix of the two.

Finland trialled the concept of a Universal Basic Income. The experiment selected 2,000 unemployed people across the country and granted them 560 euros per month without obligation to seek a job and the payment did not reduce if they found employment. The experiment ran between 2017 and 2018 and was closely followed by other countries. It aimed to identify if recipients were more likely to accept lower paid or temporary roles.

This experiment is also providing valuable insights into preparing a world for mass unemployment where governments have to help their people survive. With 560 euros per month, no one will thrive in a Scandinavian country, but everyone can survive.

THE CHALLENGES

REGULATION AND POTENTIALLY DISMANTLING TECHNOLOGY GIANTS

Now we start unveiling the dark side of digital twins. Like any major opportunity, it carries major risks. Regulators are likely to face a critical moment when they will need to decide how society should steer technology for the benefit of the masses. But they will also have to consider other factors such as the health and stability of the economy. And to do this, they will need to put safeguards in place for the use of technology.

Recent history illustrates regulators lack understanding about technology and how digital is impacting the real world.

The lack of global alignment on how regulators regulate the GAFAs (Google, Amazon, Facebook, Apple) illustrate the contradiction of interests at play.

For years, the EU has been trying to tax Google, Amazon, Facebook and Apple. These tech giants have educated their customers with outstanding digital delivery. Technology giants are now trusting huge markets like the advertising industry; they are threatening the revenue source of industries depending on it. Because those players have their regional headquarters located in fiscal havens, they can maximise profits and minimise tax which is the primary mechanism to redistribute wealth.

Tech giants have become so powerful that regulators are prosecuting them, on the ground of anti-trust regulations. They are also threatening them with astronomical fines. Google is being prosecuted for abusing market dominance in Europe's search engine market, and for anti-competitive practices and opaque pricing. The company faces billions of dollars in fines.

Amazon and Facebook went into turmoil when regulators began talking about dismantling these giants, like they did with the energy and Telco giants decades ago.

Another key challenge that regulators across the world face is privacy. The 2016 Cambridge Analytica - Facebook scandal, demonstrated to the world the role of social media platforms in the US presidential elections. That is tremendous power and a threat to strong democracies.

Regulators will have to harness privacy and leverage the concept of privacy by design and enforce it to benefit the public.

PRIVACY BY DESIGN

As we transition from an analogue to a digital world, our digital footprint dramatically increases. With respect to data, these societal and technological shifts pave the way for the potential to use centralised databases and other related infrastructure to capture, collect, store, mine and share data at an unimaginable pace and size (big data). This data could previously only be used by large organisations such as the governments and large corporates, is now available online. And it is becoming increasingly more affordable and accessible.

As citizens, consumers and professionals have slowly given away their privacy to the likes of social media, large financial institutions, utilities and governments, monitoring multiple daily transactions has become progressively more pervasive. The rise of IoT devices and 5G technology will increase that trend. With enhanced technology to capture our movements, transactions and digital profile, the asset class shifts from physical to digital. So this makes cybersecurity even more crucial. More than ever the question of privacy and customer trust is at the centre of new business models and offerings. Privacy and consumer data rights is not a new concept. It dates back to 2010. Global Data Protection Regulation - GDPR, came in to force in the EU in 2019. GDPR reflects how long it takes for regulators to keep up with technology and societal changes. GDPR introduced the notion of obtaining consent from citizens to collect their data. It also introduced the right to be forgotten and have one's digital footprint erased. GDPR also forces institutions to disclose data breaches and to consider privacy by design (PbD).

PbD is an opportunity to build trust with customers by using their data to generate value for them, not from them. PbD is an approved International Standard to preserve the future of privacy by the International Data Protection Privacy Commissioners.

By applying this PbD standard, organisations are considering privacy as a fundamental building block for designing product and service offerings that require personal data. By making privacy and customer consent the cornerstone of a value proposition, the PbD approach makes a significant contribution to building trust.

PbD revolves around the following seven key principles:

- **Proactivity not reactivity; preventive not remedial.** By being proactive with privacy, it maximises the chance to comply with foreseeable regulations that are remedial in nature.
- **Privacy as a Default setting**. Privacy by default not only protects customers but it also limits the risk of an audit by the regulators.
- **Privacy embedded into design**. PbD ensures that privacy is embedded into design and architecture. If built as a fundamental building block, privacy becomes an integral part of the system.
- **Full functionality**. By designing both privacy and security as building blocks of a system, it ensures not only that they will coexist, but they will also be mutually beneficial.
- **End-to-end security**. It is unrealistic to consider PbD without strong security practices in place. PbD ensures that data is protected throughout the lifecycle, from inception to consumption through to destruction.
- **Visibility and transparency**. PbD promotes the importance of transparency for organisations and their systems, such as data gathering practices, etc... It helps customers, stakeholders, and

communities be aware and confident about the usage of their information.

- **Respect for user privacy**. This requires organisations to genuinely have respect for their customers' privacy. Organisations can demonstrate this by establishing practices, processes and operations that are user centric. This presents a significant digital transformation challenge for organisations which have taken the approach that customers are an attribute of their products and services.

Regulators are putting safeguards in place using a combination of PbD, GDPR and the right to be forgotten, will place safeguards. They are unlikely to be enough as regulators always play catch up with societal changes.

CONCLUSION

This book was not designed to illustrate how an utopia remains an utopia until it's no longer one.

History has proven time and time again that disruption always strikes when all Necessary External Conditions (NECs) for change unite and connect. Disruption always comes fast when NECs connect with each other and scale. NECs can be through technology adoption, societal changes, compelling propositions or simply ease of use.

Well established players like Nokia and the infamous Kodak, are now case studies taught to business students to illustrate how the inability of management to embrace foreseeable change can affect the business.

On the other hand, companies like IBM and Microsoft have an extraordinary track record of embracing change and pivoting their business models to remain relevant in market.

In the early 2000s, IBM sold their hardware division to Lenovo at a point in time where they had a strong market position. At the time, they had recognised that software and services were going to be markets they were better equipped to compete in and dominate.

Microsoft and founder, Bill Gates, recognised they were late to the cloud computing race of and rapidly pivoted the organisation to compete with Google and AWS.

All the NECs for digital twins to rise are already present. Let's review the NEC categorisation across technology enablers, societal trends, emerging forms of consumption, macroeconomics and monetary policies, and emerging asset classes.

Technological enablers:

- Exponential computing power, widely available and incredibly cheap.
- Wide adoption of smartphone and AI enabled assistants like Siri, Alexa and Google Assistant, and shaping the mind for the use of emergence of digital twins.
- DL and NLP to enable the mining of minds.
- Psychology and neuroscience experts teaming up with technologists to build mind blowing virtual humans. This technology will trick our eyes and brains. For now, digital humans are designed to be recognised as virtual. This design principle could change. The rise of "deep fake" proves that with the wrong intentions our minds can be easily tricked and exploited.
- Unlimited access to information and data.

Societal trends:

- A world hyper connected, breaking geographical boundaries.
- A digital economy empowering gig workers to monetise their talent on a P2P model. This model challenges the established employment paradigm of the traditional employer-employee relationship.
- Emerging forms of trust enabling interactions and transactions in the digital economy.

Emerging forms of consumption

New forms of accessing and consuming assets, through platforms businesses and subscription models. This challenges the concept of ownership on which was based our economy.

Macroeconomics and monetary policies

Cheap access to capital due to quantitative easing policy.

Emerging asset classes

The rise of digital asset classes such as cryptocurrencies or digital reputation.

All these NECs fuel the dream that mankind can augment itself, be omnipotent and delegate low value tasks to raise their consciousness through the practice and mastery of more challenging intellectual pursuits.

For leaders or influencers of the organisation's future or if you have a duty of care to your stakeholders (shareholders, people, community), I want to ask you the following.

If you had to pivot your company to harness the power of digital twins in their emerging or foreseeable forms, what would you do? Where would you start?

How do you think your organisation is positioned against the following key internal and external attributes?

People:

- How would you assess the readiness of your leadership team? Are they ready to accept the concepts of digital twins and its impact on your organisation value chain? Are they prepared to establish a roadmap to adjust your organisation? Are they prepared to commit to and enter into a multi-year transformation program?

- How is your workforce ready to cope with having digital humans as co-workers?
- How will digital twins impact your workforce characteristics in the next 5-10 years? Which roles will be the most disrupted? Can you upskill your workforce or will there be significant redundancy?
- Do you have a plan to adapt and adjust your workforce?

Customers:

- How broadly has your customer base adopted digital? The answer might differ if you operate in the aged care industry or in social media.
- Is your current customer base likely to change?
- If yes, have you profiled them?
- Is your organisation best positioned to provide them with services and products adapted to their emerging needs?
- Is your organisation ready to target at and service your customer base digital twins'?

Shareholders:

- How can you estimate the impact on your company valuation if you do nothing?
- How would you estimate the impact on your company valuation if you were to prepare for harnessing the power of digital twins?
- How do you think you would pitch it to the investor community?

Strategic Assets:

- What strategic assets are at risk and exposed to disruption from digital twins?
- Which of your organisation's capabilities will become irrelevant? Which one can you leverage to transform?

- How do you assess the readiness of your technology stack? Is your organisation tied to legacy systems? Have you undertaken a digital transformation?

Competitors:

- How would your competitors harness the power of digital twins to put you out of business?
- Are your competitors still the same or is it a changing landscape?
- If yes, who are those emerging competitors?
- Does your business represent a core activity or an adjacency for your emerging competitors?

Partners:

- How would you map out the disruption to your value chain?
- You may be ready to adapt to a world where digital twins are at scale but are all your Tier 1/2 partners ready?
- What is your contingency plan if your key partners can't adapt?

Regulators:

- If your organisation operates in a highly regulated environment (like it is the case for financial services) what competitive advantage or disadvantage does your organisation have?
- What will be your edge to harness the power of digital twins in this environment? Will your direct competitors leverage the same edge? Will your emerging direct and indirect competitors be bound to the same rules?

Community:

- How will the community react to your organisation's decision to harness or not the power of digital twins?

The most important question of all remains the following:

- How long do you think you have left before disruption strikes?

Chances are you already ask yourself those questions but without considering the threat and opportunity that is coming.

The digital twins (R)evolution has already started. For example, it is already improving maintenance operation on heavy machinery. Insurers use it to replicate asset attributes and automate significant customer pain points in the claims value chain. Digital twins are used in engineering to design complex machines such as energy plants and planes. And in the construction industry to replicate the attributes of skyscrapers. Customer Relationship Management can use digital twins to enable digital humans to interact with real customers and act as brand ambassadors.

The next evolution of enabling mass access to the technology to replicate themselves is probably a decade away.

Hopefully, you understand I've widely used the established term "digital twins" to describe the coming disruption. In fact, what I've been talking about is closer to "digital clones". These are exact digital replicas of who you really are.

While reading you probably asked yourself what your life would look like if you could clone yourself and delegate low value, transactional and less meaningful tasks to your digital clone. How would you spend your disposable time when you have more of it? Would you focus on your business? Would you focus on your family and friends? Would you focus on your health? Would you travel the world? Would you explore a different career path while your digital clone takes care of generating income for you? Would you dedicate yourself to a charitable cause?

To stay relevant in this context most incumbents and their leaders need to start their transformation now or face the risk of becoming irrelevant.

I will conclude this book with this final quote, hoping that it will resonate with you and trigger a call to action to prepare yourself and your organisation for the disruption which has already started.

"When the wind of change starts blowing, some build walls when others build windmills." – Chinese proverb

ABOUT THE AUTHOR

John Blicq is the founder and director of Innovation Matters, a consultancy specialising in strategy development and innovation management.

John has had multiple lives and worked across Western Europe and the Asia Pacific. He started as an entrepreneur during his time in business school. After successfully selling his start-up, he worked as a management consultant and then joined a Greentech start-up/scale-up building new businesses. John has also held senior leadership roles in a number of large financial services institutions, and lectured MBA students in France and Australia.

John is a Graduate of the Australian Institute of Company Directors (GAICD) and sits on the Board of Advisors for a start-up that specialises in digital reputation enablement.

For more information about John and to get in contact:

Email: **John@innovation-matters.com.au**

www.innovation-matters.com.au

LinkedIn: **www.linkedin.com/in/john-blicq**

Medium.com: **https://john-blicq.medium.com/**

Twitter: **@johnblicq**

Made in the USA
Monee, IL
07 October 2021

79518023R00085